Anatomy
Physio.
College L
Perfect For health
Professionals &
College Students!

Fundamentals of Human
Anatomy & Physiology!

By Michael Van Sluyters

Table of Contents

Chapter 18. Developmental Anatomy and Physiology 368

Introduction to Anatomy and Physiology

The human body is a complicated machine. It consists of many parts with complex structures from cells and tissues to organs and systems.

It has different systems that are dependent on each other. They work together to make our bodies function as a single unit. In this book, you can learn about the human body's structure and its functions, particularly the following:

- Cell Anatomy and Physiology - Cells are the basic building block of every organism. Each cell has a specialized function. Learn about its parts, purpose, and how it transforms substances.

- Body Tissues - Each tissue in the human body consists of grouped cells. Our tissues are the building block of our organs and other parts of the body. It is a level between cells and a complete organ.

- Integumentary System - This organ system includes the skin, nails, hair, and glands. Learn how it protects our bodies from environmental damage.

- The Musculoskeletal System - It consists of connective tissues to support organs to function. Such as ligaments, tendons, and cartilage. These tissues give the human body the ability to move using muscles and bones.

- Central Nervous System - This system involves the brain and the spinal cord. Bones, membranes, and fluids protect these organs.

- Peripheral Nervous System - This is the system outside the brain and spinal cord. Its main role is to connect the organs and other parts to the Central Nervous System.

- Autonomic Nervous System - This system regulates involuntary functions. A few examples are digestive functions, respiration, and heartbeat.

- Endocrine System - Also known as the chemical messenger system. This system releases hormones vital for growth and function.
- Cardiovascular System Physiology - The cardiovascular system delivers oxygen, nutrients, and hormones through blood circulation. It also removes metabolic wastes from the body.

- Heart Anatomy and Physiology - The heart's function is to deliver oxygen. It gives blood to the tissues and carries away waste. The heart plays a vital role in blood circulation.

- Blood and Blood Vessel Anatomy and Physiology - This chapter focuses on how blood is circulated throughout the body via vessels, facilitating nutrients and waste exchange.

- Lymphatic and Immune System - This system works to fight against diseases. It has a network of nodes and vessels to filter antibodies.

- The Respiratory System – This is where the exchange of gases happens, providing the body the needed oxygen supply.

- Digestive System - This involves the breakdown of food in the gastrointestinal organs. This also includes nutrient absorption and excretion.

- Metabolism and Human Nutrition - This tackle on how food becomes energy. The nutrient goes into the bloodstream and gets metabolized by a specific cell.

- Urinary System Fluids, Electrolytes, and the Acid-Base System - This system eliminates waste from our body through urination. It controls the levels of fluid within our cells, as well as blood acidity.

- Reproductive System Physiology - The main purpose of the male and female reproductive organs is to reproduce. This system functions with sex hormones, fluids, and pheromones.

- Developmental Anatomy and Physiology - This covers the changes and progress in growth from fertilization. This includes embryology, fetology, and postnatal development.

- Immune System Physiology - This system works as a network of cells, tissues, and organs. It creates strategies on how to protect our bodies. The white blood cells play an important role in protection. These cells seek and fight disease-causing organisms.

- Kidney Physiology - This system consists of two bean-shaped organs called the kidneys. Its main function is to excrete waste and toxins. It also produces hormones, regulates PH, and ensure an adequate plasma.

Why Read This Book

Some students might already have the basic idea of how the human body works and moves. But, anatomy and physiology deal with it at cellular levels.

The knowledge you can gain in studying anatomy and physiology can help you in many aspects. This course can give you an insight into the human body and its functions. It covers how the body works and how to stay in its stable condition. This gives comprehensive information on what is happening inside the human body.

This is a mandatory course for someone who is in the field of medicine. Familiarity with the human body can help you decide on which action to make, particularly when a certain disease arises.

Being knowledgeable in this field can also help you understand nutrition as well as gain an understanding of possible issues (i.e. diseases). It can also help you to be aware of medical procedures during hospital admissions so you can help yourself or your loved ones mentally and physically prepare.

Medical staff who are well-informed about anatomy and physiology knows proper care. They are mindful of what causes their patient's condition and how to help them. They understand the progress of sickness and healing.

Studying anatomy and physiology can also be beneficial to your health. By knowing how organs function and get diseases, it can lead you to better lifestyle choices.

You cannot fix something that you do not understand. The organs of the human body functions as a whole; the body cannot survive if one organ is failing.

This book will be your starting point toward understanding the human body.

Chapter 1. Cell Anatomy and Physiology

The cell is the most basic and important unit of all living organisms. Its main components are water, which makes up almost 70% of the cell's weight, and biomolecules such as proteins, lipids, and amino acids, among others. These components are assembled in structural units.

Organisms living in this earth are classified either as unicellular (made up of a single cell such as bacteria and the malaria parasite) or multicellular (organized into groups of cells such as plants and animals). The amounts of cells in multicellular organisms differ from species to species. For humans, it is estimated that an average person has about 40 trillion cells that vary in size from 1 to 100 micrometers.

Cell Types

There are two major types of cells – the prokaryotic cell and the eukaryotic cell.

The Prokaryotic Cell

Bacteria and archaea are prokaryotes. Prokaryotes are one-celled organisms . They:

- Have no visible nucleus
- Are smaller and simple in structure than eukaryotic cells
- Have a plasma membrane and a cell wall covered by an outer layer called a capsule
- Have a cytoplasm that bears chromosomes
- Have no membrane-bound organelles
- Have flagella and pili that are responsible for cell movement and communication

The Eukaryotic Cell

Fungi, algae, protozoa, plants, and animals are eukaryotes. Eukaryotes are either one-celled or multicellular. They:

- Have a nucleus
- Are larger than prokaryotic cells
- Have distinct organelles arranged and organized to do specific activities
- Have cilia for cell communication
- Have flagella for mobile eukaryotes

All high forms of animals, including humans, have eukaryotic cells.

The Human Cell

The human cell is composed of three main parts: (1) the plasma membrane or the outer cell membrane, (2) the cytoplasm which is a semi-thick fluid inside the cell membrane, and (3) the nucleus.

The Plasma Membrane

The plasma membrane (also known as the cytoplasmic membrane or the cell membrane) is an elastic, porous covering that maintains the cell's shape and protects the cell from the outside environment. The plasma membrane is filled with proteins and phospholipids (fats). The phospholipids are set up in a double layer with the large protein molecules loosely attached inside and outside the surface of the membrane. Some protein molecules are specialized that they form tiny channels to regulate the entrance and the exit of fluids and nutrients

Aside from regulating the flow of substances and waste that move in and out of the cell, the plasma membrane provides shape and strength to the cell by acting like a skin that is attached to the cell's cytoskeleton. This flexible covering allows certain kinds of cells, like the red and white blood cells, to change shape when passing through veins and capillaries.

Another critical function of the plasma membrane is its ability to bind itself to the other groups of cells to form tissues.

The Cytoplasm

The cytoplasm is the clear, jelly-like substance that is inside the plasma membrane. It has three components that are critical to the life of a cell:

1. Cytosol

The cytosol is the watery part of the cytoplasm in which intercellular molecules mix, dissolve and react. Proteins, fats, carbohydrates, enzymes, and ions such as potassium, sodium, chloride, and calcium are the common molecules that exist within the fluid environment.

2. Cell Inclusions

The cell produces certain substances which float around the liquid environment of the cytoplasm. These large particles could be fat, glycogen, or melanin.

3. Organelles

As a living unit, a cell has its little organic units or "little organs" called organelles which have very distinct features and perform specific roles. Most of these organelles are invisible to the light microscope and can only be viewed under an electron microscope.

a. Ribosomes

Ribosomes are responsible for combining amino acids to form proteins. They are tiny, sphere-shaped cell organs that float freely around the cytosol or are attached to the rough endoplasmic reticulum.

b. Endoplasmic Reticulum (ER)

The ER is a group of membranes spread within the cytoplasm. They form a network of channels which connect the plasma membrane to the nucleus. There are two kinds of ER. There's the smooth ER where fats are produced and in some cells, where chemical compounds like pesticides, alcohol and cancer-causing agents are deactivated . The rough ER, on the other hand, has ribosomes stuck on its surface and is tasked to make proteins that are stored, modified and transported using its network of channels in the cytoplasm

c. Mitochondria

The mitochondria (singular – mitochondrion) are the energy generators or power producers of the cell. The organelles create energy by "burning" carbohydrate molecules like glucose. The production of energy requires oxygen with carbon dioxide as a waste product.

Energy is produced as shown by the equation:

Oxygen + Glucose = water + carbon dioxide + ENERGY

The mitochondria are rod or oval-shaped cell organs that are scattered within the cytoplasm's jelly mass. They have two membranes in which the inner one is folded for increased surface area.

Tissues and organs, like muscles, kidney, heart, and liver that need a huge amount of energy to function, have a huge number of mitochondria.

d. Golgi Apparatus

The Golgi apparatus is responsible for sorting and modifying the fats and proteins that are produced by the ER. It packs the fat and protein molecules in a membrane as vesicles for easy transport to the other parts of the cell.

The Golgi apparatus is a collection of Golgi bodies which are flat membranes piled on top of each other like a stack of plates.

e. Lysosomes

The lysosomes carry digestive enzymes that break down foreign substances like bacteria which are captured by the cell via pinocytosis or phagocytosis. The lysosomes also absorb worn-out or damaged organelles which are then recycled to make new structures for the cell.

f. Vaults

A newly discovered organelle, vaults are responsible for the delivery of the messenger RNA through the dense fluid of the cytosol to the ribosomes. Vaults are believed to be the cause in a cell's drug resistance.

Vaults are large ribonucleic particles and are three times larger than ribosomes.

g. Microfilaments and Microtubules

Microfilaments and microtubules are the thread-like structures involved in the cell's movement and ability to adjust its shape. Their task also involves the moving of chemicals and organelles inside the cell.

The microtubules are the ones which form the flagella as in the tail of the sperm and cilia which lines the respiratory tract that removes mucus and trapped dust particles.

The microtubules are also in the nucleus, forming a pair of cylindrical structures called centrioles which gather spindles needed in cell division.

The Nucleus

Visible under a light microscope, the nucleus is the most noticeable structure in a cell. The nucleus may be considered the "brain" because it is responsible for the activity and growth of the cell.

The nucleus is an oval or sphere-shaped body surrounded by a double membrane (nuclear envelope) like the plasma membrane. The membranes around the nucleus are filled with pores to allow communication between the cytoplasm and the nucleus.

Strictly speaking, the chromosomes and not the nucleus are the ones which control the cell. The chromosomes are thread-like structures that:

- Control the actions of the cell

- Carry DNA

- Are transferred to a new person when sex cells combine in sexual reproduction

- Are passed from one cell to another cell when they divide

- Vary in number from species to species

- Are constant in number in one species (e.g. humans = 46, horses = 64)

- Exist in pairs in which the members are similar in shape and length

Cell Physiology

For life to exist, it has to use molecules and do processes that follow the laws of physical science. These processes and their complex use in a cell give the basis and explanation of how a living thing could undergo growth, metabolism, and reproduction.

For materials and nutrients to pass the porous but strong plasma membrane, some processes need energy (active) while others would not need the use of energy (passive). Diffusion and osmosis are passive processes while active transport, phagocytosis, pinocytosis, and exocytosis fall under active processes.

Passive Processes

1. Diffusion

Diffusion is defined as the movement of a high concentration of liquid or air molecules to an area of zero or less concentration of molecules.

The diffusion stops when molecules are distributed equally or reach the same concentration in a closed space or container. Examples of diffusion are the smell of perfume across a room once its bottle is opened and the pouring of a teaspoon of powdered juice on a glass of water.

Within the cell, there are molecules like water, oxygen, carbon dioxide, and fats which are small enough to pass the tiny channels of the plasma membrane. Diffusion causes the small molecules to move from one side of the cell that has a high concentration, crossing the membrane, to the other side of lower concentration until both sides reach equilibrium.

For animals and humans, diffusion plays an important part in the exchange of oxygen and carbon dioxide between the blood and lungs, the movement of digested nutrients from the intestines to the blood, and the disposal of waste materials from the cell.

2. Osmosis

Osmosis is considered a special kind of diffusion in which water diffuses across a semi-permeable membrane. In osmosis, the membrane blocks molecules that are larger than water and allows only the molecules of water to pass through. The process is completely passive and does not need energy

The effect of osmosis can be seen when a dried fruit swells or a carrot gets plump once soaked in water.

The biological action of osmosis can be observed when a red blood cell is placed in water. A blood cell carries water and a lot of cellular materials within its cytoplasm. Once soaked in water, the water molecules from the outside environment move in by crossing the blood cell's plasma membrane causing the cell to swell and burst (hemolysis).

The process of osmosis is important in the moving of water from the gut to the blood and from the blood capillaries to the surrounding fluid of the cell tissues. The process also plays a critical role in the creation and elimination of urine in the kidney.

Active Processes

1. Active Transport

To keep a balance on the concentration of fluids that go inside and outside of a cell, there will be times that certain molecules, substances, or nutrients have to be transported despite their relatively big size or low amount of concentration. The process by which energy is required to force a substance of less concentration to be transported across a membrane to an area of high concentration is called active transport.

Active transport is critical in maintaining the difficult balance of concentration and exchange of sodium and potassium ions inside and outside of the cell. The process also plays an active part in the recovery of important molecules like amino acids, sodium ions, and glucose from urine.

2. Phagocytosis

Phagocytosis is the process in which the cell uses energy in moving solid particles across its plasma membrane. The process involves "cell eating" where finger-like projections from the plasma membrane encircle a solid material and slowly forces the foreign object inside the cell. The lysosomes' enzymes then destroy the alien substance.

Phagocytosis is the process used by the white blood cells in destroying harmful microorganisms and substances that try to invade the human body.

3. Pinocytosis

The mode of action of pinocytosis is similar to phagocytosis. The only difference is that foreign or potentially harmful fluid is absorbed and destroyed by the cell.

4. Exocytosis

The transport or elimination of a substance across the plasma membrane from the inside to the outside of a cell is the mode of action of exocytosis. This process is common among nerve cells and secretory cells.

Chapter 2. Body Tissues

A body tissue is a collection of similar cells that perform a specific function. Body tissues grouped together form the various organs of the body. Tissues are mainly grouped into four types: epithelial, connective, muscular, and nervous.

All cells that make up a tissue come from the fertilized egg (zygote) which over time becomes the many-celled embryo. The earliest embryonic cells can differentiate into any cell.

As cell division continues, three germ layers of cells are fashioned – the ectoderm or external layer, the mesoderm or central layer, and the endoderm or the innermost layer. Different types of tissues come from different germ layers.

Germer Layers and Resulting Organs

The body tissues form the organs, which develop from the embryo. These are the three layers of the embryo and their resultant products:

Ectoderm: epidermis (external skin layer), skin glands, some bones of the skull, the adrenal and pituitary medulla, the interior part of the mouth, the anus, and the components of the nervous system.

Mesoderm: connective tissues, cartilage, bones, blood vessel endothelium, blood, muscles, serous membranes of body cavities, synovial membranes of joints, lining of the gonads, and the kidneys.

Endoderm: some glands such as the endocrine glands, digestive glands, and the adrenal cortex, the lining of the airways, and the lining of the digestive system excluding the mouth, anal canal, and rectum.

Major Classifications of Tissues

As previously mentioned, tissues are generally classified into four:

Epithelial Tissue

Epithelial tissues help with protection, absorption, secretion, and waste elimination. They are found on the surfaces of the body (such as the epidermis or the skin's outermost layer) and within the body cavities. Sheets of tightly packed cells characterize these tissues.

This tissue type covers and protects all surfaces of the body that contact the surroundings. The skin, the digestive tract, the reproductive tract, and the airways have linings of epithelial tissue.

The epithelial tissue is kept away from others via the basal lamina, which is an extracellular matrix created from the epithelial cells. They prevent leakages, control fluid movement, and block harmful materials.

Most of the epithelium arises from the endoderm (inner layer) and the ectoderm (outer layer) during the development of the embryo. Some of it comes from the middle layer called the mesoderm.

The endothelium of blood vessels, bones, and muscles is a specialized form of epithelium; in other words, it is made of epithelial cells that changed eventually. In comparison, unchanged epithelial tissue exists as a cellular layer glued together by multiple proteins. This creates a barrier that is selectively permeable – it lets certain substances through while blocking others out.

The epithelial tissues contribute to the absorption of nutrients and fluids. They help with hormone and enzyme secretion because they are found in glands. Other than that, they secrete fluids such as saliva, sweat, and mucus.

An epithelial tissue layer has a bottom and top side. The bottom or basal side comes into contact with the cells beneath it. The top or apical side meets the interior of a hollow space in the body or the outside environment. As such, it is exposed to air or liquids, so it is designed to block them out or let select substances in.

Epithelial tissues may have special structures that help them perform their function better. For example, tissues in the intestinal walls have villi, which are minuscule finger-like structures that absorb dietary nutrients.

Because of their ability to release substances, epithelial tissues play a role in waste elimination. They trap pathogens and toxins and release them through secretions like mucus and sweat.

Kinds of Epithelium

Epithelial tissues differ in terms of structure, function, and location:

Simple squamous epithelium is a thin, flat layer of epithelial cells. This permits certain materials to penetrate the barrier through filtration and diffusion. It also secretes lubrication to reduce wear and tear of the bodily structures. Lymphatic vessels, blood vessels, the cardiac lining and the air sacs have simple squamous epithelium.

Simple cuboidal epithelium is one layer of box-like epithelial cells. This absorbs and secretes substances as well. The ducts of the glands and the tubules of the kidney have this epithelium.

Simple columnar epithelium is a single layer of tall epithelial cells. Aside from absorption, it also assists in the secretion of enzymes and mucous. Examples of simple columnar epithelium are ciliated (with hair-like structures) tissues of the uterus, uterine tubes, and bronchi, while non-ciliated tissues are located within the bladder and the digestive tract.

Pseudostratified columnar epithelium has one layer of irregularly shaped epithelial cells that give off an appearance of being stratified (having numerous layers). These secrete and move mucus via the cilia. Falsely stratified epithelium lines the upper respiratory tract and trachea.

Stratified squamous epithelium has numerous layers of thin epithelial cells. Being thick, this epithelium guards against abrasion. The mouth, esophagus, and vagina have this.

Stratified cuboidal epithelium has multiple layers of cube epithelial cells. It is a protective structure found in the salivary, sweat, and mammary glands.

Stratified columnar epithelium has several layers of elongated epithelial cells. It protects the structure it covers and secretes substances and wastes. Some of the glands and the male urethra have stratified columnar epithelium.

Transitional epithelium has several layers of odd-shaped epithelial cells. Their design permits the expansion of urinary organs such as the urethra, ureters, and bladder.

Connective Tissues

Connective tissues serve as the binding material of organs and cells. These are fibrous and have an extracellular matrix consisting of inorganic substances. The protein fibers in connective tissues are called collagen and elastin – these are produced by fibroblast cells.

The matrix of these tissues may be in solid, jelly, or liquid form. For instance, the blood's extracellular matrix is plasma fluid, while the bones have a rigid matrix built from minerals and collagen fibers.

The most common kind of connective tissue is loose connective tissue. It links muscles to epithelial cells and supports blood vessels and internal organs.

Fibrous or dense connective tissues comprise the ligaments and tendons. These attach bones to other bones and to the muscles that control them.

Some examples of connective tissue that have developed specialized forms are cartilage, bone, body fat (adipose tissue), and blood.

Muscle Tissues

Muscle tissues form out of the mesoderm layer of the embryo. These enable movement and keep the body upright. They help with moving food through the gastrointestinal tract and to pump blood all throughout the circulatory system.

The muscle cells (also known as muscle fibers) have myosin and actin proteins that enable their contraction. They are grouped into three major kinds: skeletal, cardiac, and smooth.

The skeletal muscle is sometimes called as striped muscle because of its appearance. People know this muscle type the most. Skeletal muscles are connected to the bones through the tendons. Some examples of this muscle type are triceps, biceps, and quadriceps.

Cardiac muscles exist only in the heart's walls. These are also striated or striped like skeletal muscles, but they cannot be controlled voluntarily. The muscular fibers are connected by intercalated disks that make them contract together.

Smooth muscles make up the walls of internal structures such as the uterus, digestive tract, urinary bladder, and the blood vessels. They are not striated or striped and they move on their own (without conscious control).

Nervous Tissue

Nervous tissues develop mostly from the ectoderm of the embryo. The excitable tissues of the nervous system generate and propagate electrochemical messages; these signals direct the varied activities within the body.

Nervous system tissues sense internal and external stimuli, process information, and transmit it to where it is needed. They have two major cell types: nerve cells (a.k.a. neurons) and neuroglia.

Neurons are the nervous system's basic unit. They create electrical signals (action potentials/conducted nerve impulses) that are sent swiftly across great distances. These nervous tissues are found in the nerves, spinal cord, and brain.

The neuroglia support the function of neurons. They prevent infection of the nervous system, guide the flow of neurotransmitters, and regulate ion concentration within the cells.

Tissue Membranes

Tissue membranes are thin sheets of cells that perform the following functions:

- Cover the external surfaces of the body
- Line the interior of the body cavity
- Serve as lining to vessels
- Cushion a joint cavity

They are classified into two types based on what kind of tissues they are made of – epithelial tissue membranes if they are composed of epithelial tissues, and connective tissue membranes for connective tissues.

Epithelial membranes include the cutaneous, serous, and mucous membranes.

Connective tissue membranes are found near joints as synovial membranes.

Epithelial Membranes

An epithelial membrane has a layer of epithelial tissue linked to another layer of connective tissue. It serves as a protective barrier or a lubricating medium.

Mucous Membranes

A mucous membrane or mucosa lines a hollow passage or body cavity that is exposed to the external environment. As such, they are present in the respiratory, reproductive, digestive, and excretory tracts.

Glandular tissue and uniglandular cells produce mucus, which creates a protective coating on the epithelial layer.

Serous Membranes

A serous membrane coats the body cavities that are unexposed to the surroundings. Epithelium cells secrete serous fluid to provide lubrication and reduce organ friction.

Three serous membranes line the thoracic cavity; two coat the lungs, and one encapsulates the heart as the pericardium. The peritoneum blankets the peritoneal cavity, enclosing the abdominal organs and branches out into two sheets of mesenteries that hold the digestive organs.

Cutaneous Membranes

A cutaneous membrane has multiple layers of connective and epithelial tissues. The top surface of this membrane faces the environment, so it is covered with a protective layer of keratinized cells. The skin is a form of a cutaneous membrane.

Synovial Membranes

Synovial membranes have connective tissue membranes that lubricate the joints to improve mobility.

In summary, collections of similar cells are called as body tissues, while the aggregations of similar tissues form the body's organs. These tissues are classified into four main types: epithelial, connective, muscular, and nervous.

The bodily tissues and the cells that compose them originate from the three layers of the embryo: the inner layer called endoderm, the middle layer or mesoderm, and the outer layer that is the ectoderm. The epithelial tissues spring forth from all of these layers, while the muscle tissues develop from the mesoderm and the nervous tissues are based from the ectoderm.

Epithelial tissues serve as the covering of surfaces and they control how substances move through and along them. Connective tissues bind and support the structures of the body. Muscle tissues enable movement whether voluntary and involuntary, while nervous tissues serve as the communication portals of information.

There are three types of muscles: skeletal, cardiac, and smooth. Skeletal muscles are linked to bones and entry points to enable voluntary motion. Cardiac muscles operate automatically to facilitate blood flow. Smooth muscles are positioned on the walls of passageways and organs to regulate respiration, digestion, and other bodily processes.

Tissue membranes can have epithelial and connective tissues. They protect surfaces and interiors from environmental elements and pathogens. Epithelial membranes include the mucous and cutaneous membranes found on external surfaces of the body, and the serous membranes within the body cavities. Synovial membranes give lubrication to joints.

Chapter 3. The Integumentary System

The Integumentary System, also called the integument, is the external covering of the body. It consists of the skin, hair, nails, and exocrine glands. It accounts for about 16% of total body weight and has a surface area of 20 square feet on an average adult person. It is also considered the largest organ in the body.

Functions of the Integumentary System

The skin and its features perform a variety of vital functions in the body, most of which are protective functions.

1.   *Protection*. The skin acts as the first line of defense of the immune system. It acts as a protective barrier between the internal body

and the external environment therefore filtering any organism that come in contact with the body.

2. *Thermoregulation.* Body temperature regulation is primarily regulated by an area in the brain called the hypothalamus. It is responsible for opening and closing the sweat glands, as well and contracting muscles. The integumentary system aides in thermoregulation through the sympathetic nervous system, the central nervous system division involved in fight-or-flight responses.

3. *Sensory function.* Various sensations are felt in the body due to the presence of sensory receptors.

4. *Vitamin D Synthesis.* During exposure to sunlight, ultraviolet (UV) radiation penetrates into the epidermis therefore producing vitamin D3, also known as

cholecalciferol. This vitamin is essential for normal absorption of phosphorus and calcium which is vital in producing and maintaining healthy bones. In addition, vitamin D is also essential for general immunity against viral, bacterial, and fungal infections.

Structure of the Skin

The skin is made up of multiple layers of cells and tissues which are attached to underlying connective tissues. Numerous sensory, autonomic, and sympathetic nerve fibers which transmits signals to the brain are also present on these layers.

The skin is composed of two layers: the epidermis, and the dermis. These two layers are made up of dense, irregular connective tissue that houses blood vessels, sweat glands, hair follicles, and other structures.

Epidermis

The epidermis is the most superficial layer of the skin. It is composed of keratinized, stratified, squamous epithelium and it has no blood supply (avascular). All layers of the epidermis are rich in keratin, an intracellular fibrous protein responsible for the hardening of the skin, nails, and hair. This protein is manufactured and stored inside small cells called keratinocytes.

The skin has four or five sub layers, depending on its location, namely stratum corneum, stratum lucidum, stratum granulosum, stratum spinosum, and stratum basale.

a. Stratum Corneum

The stratum corneum is the most superficial layer of the skin and is the layer exposed to the external environment. It is 20 to 30 layers thick, and is responsible for about three-quarters of the epidermis' thickness.

On average, a person sheds about 18 kg (40 lbs.) of stratum corneum in a lifetime. A "new" epidermis is formed every 25 to 45 days. This layer is filled with shingle-like dead cells called cornified or horny cells, commonly filled with keratin, a structural protein that makes up hair, nails, skin, and other accessory structures. This layer functions by helping the body resist physical, chemical, and biological assaults. It also helps prevent water loss.

b. Stratum Lucidum

The second most superficial layer, the stratum lucidum, is an epidermal layer that is only found on selected skin regions. It only occurs on areas where the skin is hairless and extra thick. Such regions are the palms of the hands and soles of the feet. The keratinocytes that compose this layer are dead and flattened. This layer is also filled with eleidin, a clear, protein rich lipid that gives these celled a lucid, transparent appearance. Eleidin also provides a barrier to water.

c. Stratum Granulosum

The stratum granulosum has a grainy appearance due to further changes in keratinocytes as they are pushed from the stratum spinosum. Its cells are 3 to 5 laycrs deep containing keratohyalin granules which are filed with histidine- and cysteine-rich proteins that bind keratin filaments together. Concomitantly, it forms a hydrophobic lipid envelope responsible for the skin's barrier properties.

d. Stratum Spinosum

The stratum spinosum, also called the spinous layer or the prickle cell layer, has a shiny appearance due to the protruding cell processes that result to the cells forming connections via a structure called a desmosome. Desmosomes are responsible for interlocking and strengthening the bond between cells. It has 8 to 10 layers of keratinocytes as a result of stratum basale's cell division. Intermingled among the keratinocytes of this layer is a dendritic cell known as Langerhans cell. This functions as a macrophage by engulfing foreign particles, damaged calls, and bacteria. The stratum spinosum functions by making the skin waterproof due to the synthesis of keratin. It also prevents water loss from the body.

e. Stratum Basale

The stratum basale, also known as stratum germinativum, is the deepest epidermal layer. It is a single layer of cuboidal keratinocytes. This basal layer is the most adequately nourished as it lies closest to the dermis.

t is attached to a layer in the dermis called basal lamina. The cells in stratum basale bond to the dermis through intertwining collagen fibers called basement membrane. On the superficial portion, a finger-like projection called dermal papilla is found which increases the strength of the epidermis and the dermis.

Dermis

The dermis, also called corium, is the middle layer of the skin. It is a strong, stretchy envelope that helps bind the body together. This layer is made up of connective tissue and has two layers – the **papillary layer** and the **reticular layer**. Like the epidermis, its thickness varies on its location.

a. Papillary Layer

The papillary layer is the superficial dermal region and is composed of areolar, irregular connective tissue. It has small, peg-like projections called dermal papillae that contains capillary loops, which provides nutrients to the epidermis. It also houses pain receptors (free nerve endings), and touch receptors.

On the palms of the hands and soles of the feet, these papillae are arranged in a looped, ridged pattern to increase friction and enhance the gripping.

b. Reticular Layer

The reticular layer is the deepest layer of the skin and is composed of dense, irregular connective tissue. Blood vessels, sweat and oil glands, and deep pressure receptors (lamellar corpuscles) are also found on this layer.

The dermis also contains collagen fibers which are responsible for the tough structure of the dermis. It also attracts and binds water to help keep the skin hydrated. Lastly, elastic fibers could also be found here. From the name itself, elasticity of the skin comes from this protein. However, as we age, the collagen and elastic fibers decrease, therefore giving a sagged, wrinkled appearance of the skin.

Due to the dermis' abundance of blood supply, it also contributes in maintaining the body's homeostasis. When the body's temperature increases above normal, the capillaries become swollen with heated blood and the skin becomes warm and erythematous. This allows the body heat to radiate from the skin surface in order to restore normal temperature.

Appendages of the Skin

These include cutaneous glands, hair and hair follicles, and nails. Each of these appendages play a unique role in maintaining the body's homeostasis.

Cutaneous Glands

Cutaneous glands are glands that release secretions into the skin via ducts. These glands fall into two major groups: the **sebaceous glands** and the **sudoriferous glands**. These glands are found in the dermis but are produced by the stratum basale.

The sebaceous glands, or oil glands, are found all over the body. The only parts of the body without this gland are the palms of the hands, and soles of the feet. The product of this gland is called sebum, a mixture of fragmented cells and oily substances. Sebum is a lubricant that helps keep the skin moisturized and prevents hair from becoming dry and brittle. It also has chemicals that kill bacteria to avoid skin infection. During adolescence, the sebaceous glands are very active hence androgen (male sex hormones) production is increased in both sexes.

The sudoriferous glands, also called sweat glands, are widely distributed to the skin. There are more than 2.5 million sweat glands on an average adult person. The two subtypes of this gland are eccrine and apocrine.

Eccrine glands are more numerous and found all over the body. They produce sweat, a clear secretion that is primarily made of water, vitamin C, lactic acid, some salts (sodium chloride), and traces of metabolic waste such as ammonia, urea, and uric acid. It is acidic with a pH of 4 to 6, a characteristic that helps inhibit bacteria. These glands are also efficient in regulating the body's heat. It has nerve endings that causes them to secrete sweat when the environmental temperature or body temperature is too high.

Apocrine glands are largely confined to the axillary and genital areas of the body. These are usually larger than eccrine glands, and their ducts empty into hair follicles. Its secretions contain fatty acids and proteins, and sometimes may have a milky or yellowish color. These glands begin to function during puberty due to the influence of androgen.

Hair and Hair Follicles

Hair is a flexible epithelial structure. A part of it is enclosed in the hair follicle called the root; while the part that projects from the surface of the scalp is called the shaft.

Each hair strand has a central core called the medulla which consists of large cells and air spaces. It is enclosed by a bulky cortex layer that is made up of several layers of flattened cells. The cortex, in turn, is encapsulated by a cuticle, a single layer of cells that overlap one another.

The cuticle is the most heavily keratinized section and it helps keep the hairs apart. It also provides strength and keeps the inner hair layers tightly compacted. Hair pigment have several varieties depending on the amount of melanin. It also comes in different shapes and sizes. They are short and stiff on the eyebrows, smooth and silky on the head, usually nearly invisible almost everywhere else.

Hair follicles are compound structures. The inner epithelial root sheath is made up of epithelial tissue and it forms the hair. The outer fibrous sheath on the other hand is made up of dermal connective tissue. It supplies blood vessels to the epidermal portion and reinforces it. Small bands of smooth muscle cells called arrector pili are also present. When experiencing "goosebumps", these muscles contract and hair is pulled upright.

The hair growth cycle has three distinct stages namely the anagen phase, catagen phase, and telogen phase. Anagen phase is the most initial stage wherein the hair grows half an inch within a month. This usually lasts from3 to 7 years, and the average full-length hair is 18 to 30 inches. Catagen phase is also known as transitional phase as this only lasts for approximately 10 days. The last stage called telogen phase is a resting phase where the hair falls out. The normal amount of hair fall in a day is 100 strands. The follicles remain inactive for 3 months and later on goes back to anagen phase.

Nails

These are scale-like modifications of the epidermis which in other animals may appear as claws. Nails are transparent and nearly colorless, but has a pinkish appearance due to the rich blood supply in the underlying dermis.

Each nail consists of a free edge, a body (visible attached portion), and a root (embedded in the skin). It also has nail folds which are the borders of the nail overlapped by folds of skin. The edge of the proximal nail fold is called the cuticle.

The nail bed is an extension of the stratum basale. It has a thickened proximal area called the nail matrix that is responsible for nail growth. As nail cells are produced, they become heavily keratinized and die hence nails are mostly nonliving material like hairs

Fingernails grow faster than toenails with a production rate of $1/10^{th}$ every month. New nail cells are constantly produced and form the nail root. These cells are hardened as keratin added, and has a healthy pink color due to the blood vessels on the nail bed.

Development of Skin Color

There are three pigments that contribute to the skin's color namely: melanin, carotene, and hemoglobin.

Skin color primarily comes from melanin, this pigment produces a yellow, reddish brown, or black color to the skin. Upon exposure to sun, melanocytes are stimulated to produce more melanin pigment, resulting to tanning of skin. As melanin is produced, the melanocytes accumulate in a cytoplasm in membrane-bound granules called melanosomes. These granules then attach to nearby keratinocytes. Inside the keratinocytes, the melanin forms a pigment umbrella over their nuclei and shields its genetic material (DNA) from the UV rays. People with plenty of melanin have brown-toned skin, whereas people with are light skinned.

Carotene is an orange-yellow pigment found in carrots and green leafy vegetables. It is most commonly deposited into the stratum corneum and subcutaneous tissue. People who eat carotene-rich foods tend to have a yellow-orange cast on their skin.

Hemoglobin is the pigment found in red blood cells. The amount of oxygen-rich hemoglobin in the blood vessels is responsible for a rosy glow of the skin due to its crimson-colored appearance.

Temperature Homeostasis

Heat and cold receptors can be found in the skin. When the body temperature is elevated, the hypothalamus sends nerve signals to the sudoriferous glands, causing it to release 1 to 2 liters of water per hour resulting to cooling of the body. The hypothalamus also sends signals for dilation of blood vessels to cause heat to convect away from the skin's surface. On the other hand, the sweat glands constrict and sweat production decreases when body temperature falls. If it continues to decrease, and increase in the body's metabolic rate occurs by and this manifests in the body shivering.

Chapter 4. The Musculoskeletal System

The musculoskeletal system has two components: the muscular system and the skeletal system. The Musculoskeletal system is the biological system of humans that is responsible for producing movements, which is why it is also called the locomotor system, and traditionally, the activity system.

The Musculoskeletal System interact with other body systems in various ways. The circulatory system interacts with the musculoskeletal system by providing proper nutrients to the muscles for efficient movements. It also takes away waste products. Another function it has is the regulation of hormones needed in producing movements.

It is also in a way connected to the digestive system. The very first step in digestion is chewing, and this requires movement of the jaw. Such movements are produced by the musculoskeletal system. Other than that, the insides of the organs where food pass through are made of smooth muscles. The sphincters that control the openings of these organs are made of smooth muscles as well.

The respiratory system's most important structure is the diaphragm, a large muscle found in between the lungs and intestines. Upon movement of the diaphragm, air enters and exits our body. This is possible due to the smooth muscles that control the said organ.

The Muscular System

The muscular system, while at times used interchangeably with "musculoskeletal system" is actually the organ system that consists of the muscles that make up the body particularly skeletal muscles, smooth muscles, and cardiac muscles.

Functions of the Muscular System

The muscular systems has four major functions: movement production, joint stabilization, maintaining posture and body position, and heat generation.

The muscular system is responsible for locomotion and mobility of the body. It enables the body to quickly respond to changes in the external environment.

All skeletal muscles are connected to bones and joints, therefore causing movements and stabilization. The tendons of the muscles play a role in stabilizing joints with poorly fitting articular surfaces.

The muscles of the body are functioning continuously to help keep our bodies erect. It also helps balance or counteract the pull of gravity.

Muscle activity also produces body heat as its by-product. This is vital in maintaining normal body temperature.

Muscles have significant functions in internal organs as well. Some of its functions include dilation and constriction of our pupils, regulation of the valves that regulate our internal organs, and protection of vital organs among others.

Types of Muscle Tissue

There are three different types of muscle tissue – skeletal, smooth, and cardiac. These differ by their location, cell structure, and the stimulation it needs to contract.

CHARACTERISTIC	SKELETAL MUSCLE	CARDIAC MUSCLE
Location	Attached to bones or skin	Wall of the h
Cell shape and appearance	Single, very long, cylindrical, multinucleate cells with striations	Uninucleate, striated
Connective tissue components	Epimysium, perimysium, endomysium	Endomysium
Regulation of contraction	Voluntary	Involuntary
Speed of contraction	Slow to fast	Slow
Rhythmic contraction	No	Yes

Skeletal Muscle

The skeletal muscle is the type that creates movement in the body. There are more than 600 skeletal muscles and it makes up about 40% of a person's body weight. This type is anchored to the bones via tendons to put skeletal movements in effect such as locomotion. It is called striated because of its longitudinally striped appearance.

Most skeletal muscles are firmly attached to bones via tendons. These are made up of dense regular connective tissue but are very thick and strong. This structure is a tough band found at the end of muscles.

Muscles have layers as well, most of which are mostly made up of dense connective tissues. Below are the layers from the most superficial to the deepest.

- Fascia. This is a layer of thick connective tissues that covers the entirety of a muscle.
- Epimysium. A single muscle is enclosed by its most superficial layer, the epimysium. This is made of dense connective tissue and extends beyond the fleshy part of the muscle where it creates a flat sheet-like aponeurosis called tendon.
- Perimysium. This collagen-filled layer wraps a bundle of muscle fibers called fascicles. Located in this layer are the blood vessels and nerves that supply the muscle.
- Endomysium. This is the covering of individual muscle fibers. It also contains capillaries, lymphatics, and nerves.
- Sarcolemma. This is the cell membrane the encloses each muscle cell, also known as muscle fiber.

- Sarcomere. This is the basic unit of the muscle tissue. The sarcomere has two proteins: thin filaments and thick filaments. Myosin is a protein that bonds thick filaments together. It is the major protein that causes muscle contraction. Thin filaments, on the other hand, are made up of three proteins: actin, tropomyosin, and troponin. Actin is the most abundant among the three as it contains myosin-binding sites. This allows myosin to bind with actin during muscle contraction. Tropomyosin is a long protein fiber that wraps around the actin. It covers the binding sites of the proteins. Troponin is another protein bound to tropomyosin. This helps move tropomyosin away from myosin binding sites during muscle contraction.

Subtypes

The skeletal muscle has several subtypes:

1. Type I. Also known as low oxidative, slow twitch, or "red" muscle. It is dense in characteristic and is rich in mitochondria and myoglobin, giving it a red appearance. It is also rich in oxygen and can sustain aerobic activity.

2. Type II. Also known as fast twitch, it has two major kinds:

- Type IIa. This muscle type is similar to slow twitch muscle – it is aerobic, has a red appearance, and is rich in mitochondria and capillaries.

- Type IIb. This muscle type in anaerobic, with a white appearance and less dense mitochondria and myoglobin. Small animals mostly have this type of muscle fiber hence the pale appearance of their meat.

Fascicles

As mentioned, skeletal muscles are bundles of fascicles. These fascicles are arranged in certain directions to accommodate the location and movement a certain muscle is responsible for. Some common patterns are convergent, parallel, circular, fusiform, pennate, bipennate, unipennate, and multipennate.

- Convergent pattern. This is the pattern where the fibers of the muscle direct towards the tendon. It usually forms a triangular shape such as the pectoralis major.
- Parallel pattern and Fusiform pattern. Muscles with this pattern usually run the length parallel to the long axis of the muscle such as the sartorius. Fusiform pattern is a modification wherein the

muscle is longitudinal but has an expanded midsection such as the biceps brachii.

- Circular pattern. From the name itself, circular pattern is arranged in rings. These are commonly found in external body openings. An example is the orbicularis oculi.

- Pennate pattern. Pennate, or feather, has its short fascicles attach to the central tendon in an oblique way. The term unipennate means the muscle only insert into one side of the muscle; bipennate means the fascicles insert to opposite sides of the tendon; and multipennate means the fascicles insert from several different sides.

Smooth Muscle

Smooth muscles, or visceral muscles, have no striations and are involuntary, meaning they cannot be consciously controlled. This type is mainly found in the walls of hollow, visceral organs such as respiratory passages, stomach, and urinary bladder. These muscles are controlled directly by the autonomic nervous system.

Cardiac Muscle

Cardiac muscle is unique because it can only be found in one organ of the body – the heart. This forms the bulk of the heart walls and is responsible for pumping blood throughout the body. These tissues also act as the pacemaker of the heart. Just like smooth muscles, its movements are involuntary and are controlled by the sinus node with its signals coming from the autonomic nervous system.

Cardiac and skeletal muscles are striated and have highly regular arrangement of bundles. These are usually used in short, intense bursts. Smooth muscle on the other hand has neither characteristic; it has the ability to sustain longer or near-permanent contractions. For most muscles, contraction occurs as a result of conscious effort. The brain sends signals which pass through the nervous system, therefore producing an effective contraction. However, some muscles such as cardiac cannot be voluntarily controlled. This is called autonomic.

How Skeletal Muscles' Names Indicate Information

Muscles are named based on several factors including their location, direction, origin and insertion, size, number of origins, and action.

- Location. Some muscles are named by their anatomic location. For example, the rectus abdominis is named such because it is found in the abdominal region.

- Direction of the muscle. There is an imaginary line in the midline of the body that is used as a reference for muscle direction. When a muscle has the term *rectus*, its fibers run parallel to the imaginary line. For example, the rectus abdominis runs parallel to the body's midline. If the term *oblique* is in a muscle's name, it indicates that the muscle fibers are slanted. For example, the external obliques branch out laterally at the abdominal part.

- Origin and Insertion. Some muscles are named by their connection to a stabilizer bone (origin) and a moving bone (insertion). An example is the sternocleidomastoid muscle, which is attached to the sternum, clavicle, and mastoid process.

- Size. Muscles with terms such as *maximus* (largest) and *minimus* (smallest) indicate the size of the muscle.
- Number of origins. Some muscles connect to two or more bones, therefore having more than one origin. For example, the terms *quadriceps, triceps,* and *biceps* have four, three, and two origins respectively.
- Action of the muscle. Muscles could also be named by the action they perform. An example is the extensor digitorum which acts as an extensor.

Muscle Interactions

Muscles could only work by a pulling force hence they need to work together or in pairs.

Below are the categories on the different roles muscles play during movement.

- Agonist. These muscles are also known as prime movers. It is the main muscle responsible for initiating and maintaining a movement. For example, the biceps brachii is the agonist muscle during pulling up the forearm towards the shoulder (elbow flexion).

- Antagonist. An antagonist is responsible for opposing the agonist. This acts by stretching or lengthening the opposite muscle when the agonist contracts or shortens. Example, during flexion of the elbow, the triceps brachii acts as the antagonist.

- Synergist. These muscles act around the moving joint to aide the agonist during muscle movement. For example, the brachioradialis and brachialis muscles contract during elbow flexion to help the biceps brachii.

- Fixator. From the name itself, it fixates or stabilizes the origin of the agonist so that muscle contraction is more efficient. It is also responsible for preventing unnecessary movement. With the same example, the abdominals and shoulder muscles contract during elbow flexion to avoid unnecessary motions.

Types of Body Movements

All skeletal muscles, with a few exceptions, cross at least one joint thus permitting motion. These structures have two points, the origin and the insertion. An origin is the point of the muscle that acts as an anchor point and the insertion is attached to the movable bone. Upon muscle contraction, the insertion moves towards the origin. It always works this way as skeletal muscles can only pull; they never push. Generally, body motion occurs when muscles contract across joints. Below are the most common types of body movements:

- Flexion. This motion decreases the angle of a joint and brings two bones together. This generally occurs in the sagittal plane, and is common in hinge joints (such as the elbow joint) and ball-and-socket joints (such as the hip joint).

- Extension. This motion is the opposite of flexion wherein the angle between two bones increase such as straightening the elbow or the knee. In cases where extension exceeds 180°, it is termed as hyperextension.

- Abduction. Moving a limb away from the median plane is called abduction. It occurs generally on the frontal plane. An example is fanning out the fingers or toes.

- Adduction. Adduction is the opposite of abduction. It is the movement of a limb towards the median plane. Moving your upper limb towards your thorax is an example of adduction.

- Rotation. Rotation is movement of a bone around its longitudinal axis. An example is shaking your head "no", wherein the atlas moves around the dens of the axis.

- Circumduction. This refers to a conical movement of a body part. Circumduction is a sequential combination of flexion, adduction, extension, and abduction at a single joint. This is common in multi-axial ball-and-socket joints such as the shoulder joint.

- Elevation and Depression. Moving a limb or a body part in a superior direction is termed elevation whereas moving it in an inferior direction is termed depression. An example is moving the apex of the shoulder upward (elevation) and downward (depression) using the trapezius muscle.

Aside from these movements, there are other motions that do not fit into any of the previous categories. The following occur in a few joints only:

- Supination and pronation. Supination, meaning turning backward, and pronation, meaning turning forward, refer to the movement of the radius on the ulna. Supination is the movement where the forearm rotates laterally, while pronation is the movement of the forearm medially.

- Dorsiflexion and plantarflexion. These two motions solely occur at the ankle. Dorsiflexion refers to the upward movement of the foot at the ankle joint, whereas plantarflexion is the downward movement of the foot at the ankle joint.

- Inversion and eversion. Just like dorsiflexion and plantarflexion, these two motions only occur at the ankle joint. Inversion refers to turning the sole medially, and eversion refers to turning the sole laterally.

- Opposition. This is the action wherein the thumb is moved to touch the tips of other fingers. This is needed during grasping objects. Opposition is commonly done in 1st metacarpal joint.

The Sliding Filament Theory

The Sliding Filament Theory explains the mechanism as to how muscles contract to produce force. Muscle fibers are made up of myofibrils, which comprise of sarcomeres. These sarcomeres are filled with proteins actin (thin filament) and myosin (thick filament). When a muscle contracts, the sarcomere needs to shorten. Actin and myosin are pulled together towards the center of the sarcomere until these filaments overlap. While the sarcomere shortens, the filaments' length remains the same.

A sarcomere is defined by the distance of two consecutive Z discs. When a muscle contracts, the distance between these Z discs decrease. The H zone, also shortens continuously and the actin and myosin filaments overlap thus when the muscle is in full contraction, the H zone is no longer visible. Another part of the sarcomere, the I band, only contains actin. The A band does not shorten but they move closer during contraction.

Thin filaments are pulled towards the center by the thick filaments until maximal contraction is obtained. Note that the actin and myosin do not shorten, they merely slide past each other to decrease the size of the sarcomere.

Energy Supply for Muscle Contraction

The muscle needs calcium (Ca++) ions and ATP (Adenosine Triphosphate to contract. However, very little amount of the ATP is stored in the body so they body has three different pathways to regenerate ATP.

1. Aerobic pathway

About 95% of the ATP is used for muscle activity at rest and during light exercise.

Aerobic respiration occurs wherein a series of metabolic pathways that use oxygen occurs in the mitochondria. Collectively, these pathways are called oxidative phosphorylation. During aerobic respiration, glucose is broken down into water and carbon dioxide. Some of the energy released attach to bonds of ATP molecules. But despite a large amount of ATP during aerobic respiration, it is fairly slow and requires a continuous supply of oxygen and nutrients.

2. Anaerobic glycolysis and lactic acid formation.

The first step in breaking down glucose is termed glycolysis and does not require oxygen hence it is anaerobic. During glycolysis, glucose is broken down into pyruvic avid and small amounts of energy produced during this process attach to ATP bonds. As long as oxygen is present, pyruvic acid continues to work by entering the pathways in the mitochondria to further produce ATP. Pyruvic acid generated during glycolysis is then converted to lactic acid. This is termed as anaerobic glycolysis. This procedure gives about 5% ATP only. However, it is 2.5x faster than aerobic, and can provide ATP within 20 to 30 seconds of strenuous muscle activity.

3. Direct Phosphorylation.

Creatine phosphate (CP) is a high energy molecule that is found in muscle fibers only. Once ATP depletes, CP and Adenosine Diphosphate (AD) interact and regenerate more ATP within 15 seconds.

Muscle Contraction and Relaxation

Skeletal muscle fibers need to be stimulated by nerve impulses in order to contract. One motor neuron has the ability to stimulate a few muscle fibers or hundreds of them, depending on the particular muscle and the work it does.

The contraction of an individual muscle fiber begins with Acetylcholine, a neurotransmitter that is responsible for sending signals towards the motor neuron that innervates muscles.

Once a signal is sent, the local membrane of the fiber depolarizes as positively charged Sodium ions (Na+) enter the membrane, triggering an action potential. This spreads through the entire membrane therefore depolarizing it including the t-tubules. As a result, calcium (Ca++) ions are released from the sarcoplasmic reticulum where they are stored. The Ca++ ions then initiates contraction which is sustained by ATP (Adenosine Triphospate). As long as the Ca++ ions remain binded to troponin, which keeps he actin-binding sites unshielded, and as long as ATP drives the crossbridge cycle, the muscle continues to contract.

Muscle contraction ends and relaxation begins when the signals from the motor neuron end. The sarcolemma and t-tubules are then repolarized and the voltage-gated calcium channels are closes. Ca++ ions are pumped back into the sarcoplasmic reticulum and the tropomyosin is shielded against the binding sites on the actin strands. Muscle relaxation also happens when it runs out of ATP and becomes fatigued.

Muscle Fatigue and Oxygen Deficit

Muscle fatigue occurs when it is unable to contract despite being stimulated. This occurs due to continuously performing a strenuous activity for a prolonged period. The factors for muscle fatigue is not fully known but t is believed to be connected with an imbalance in the ions in the body which results to problems in the neuromuscular junction.

Oxygen deficit is also found to contribute in muscle fatigue. This happens when the person is unable to take in oxygen fast enough to supply the muscles. When muscles lack oxygen to perform aerobic respiration, lactic acid accumulates in the muscle via the anaerobic pathway. The muscle's ATP then starts to decrease and ionic imbalance occurs.

Types of Muscle Contraction

There are different ways in which muscles contract. Muscle contraction does not always mean it shortens as muscle lengthening is also considered contraction. Below are the types of muscle contraction.

1.   Isometric Contraction. During isometric contraction, the muscle does not shorten. The

muscle tension is maintained but the length changes.

a.   Concentric Contraction. This type of muscle contraction occurs when the muscles actively shorten while generating force, thus overcoming resistance. In this type, the force generated by the muscle is less than the muscles maximum capacity. For example, during performing biceps curl, the biceps brachii concentrically contracts.

b.   Eccentric Contraction. This is a type of contraction wherein elongation of the muscle occurs while still generating force. This can be either voluntary or involuntary. Using the same example, the triceps brachii eccentrically contracts during biceps curl as it lengthens to accommodate the movement.

2.   Isotonic Contraction. Isotonic, meaning "same tone", is a type of muscle contraction

where movement occurs, but the muscle tone stays the same.

3.   Isokinetic Contraction. In this type, the tension and length changes but the speed of contraction is constant.

4.   Yielding. Yielding contraction happens when a muscle contraction is opposed by weight or resistance. For example, holding a heavy weight without raising or lowering the object.

5.   Overcoming. This type occurs when muscle contraction is opposed by an immovable object. For example, pushing against a wall.

The Skeletal System

The skeletal system includes all of the bones, cartilages, ligaments, and joints of the human body. It makes up 20% of a person's body weight.

The bone is an osseous tissue made up of hard, dense connective tissue that acts as the support structure of the body.

There are two main categories of bones: the axial skeleton and the appendicular skeleton. The axial skeleton consists of the skull which protects the brain and facial structure, the spine or vertebral column which surrounds and protects the spinal cord, and the thoracic or rib cage which surrounds and protects the organs within the chest. The appendicular skeleton, on the other hand, consists of the pectoral girdle, the pelvic girdle, and the upper and lower limbs.

Functions of the Skeletal System

The skeletal system has 5 major functions: it supports the body, facilitates movement, provides protection, produces blood cells, and acts as storage for fats and minerals.

1.   Supports the body. Bones provide support by serving as a point of attachment for the muscles.

2.   Facilitates movement. It also transmits forces when muscle contraction occurs. Mechanically, bones act as levers and joints serve as fulcrums, which will be discussed further on this chapter.

3.   Serves as protection. The skeletal system provides protection by covering and surrounding the internal organs to avoid injury. For example, the skull houses the brain to avoid injuries.

4.   Produces blood cells. Termed as hematopoiesis, blood production occurs in the bone, specifically the bone marrow. This is where all types of blood cells (red blood cells, white blood cells, and platelets) are produced.

5.   Acts as storage for fat and minerals. On a metabolic level, bones act as a reservoir for several minerals, especially calcium and phosphorus. These play an important role in muscle contraction and flow of ions in the body.

Structure and Layers of the Bone

A bone has two major parts: the diaphysis and the epiphysis. The diaphysis is the tubular shaft that runs between the proximal and distal ends of the bone. It has a hollow region filled called the medullary cavity which is filled with yellow marrow, where some of our blood is produced.

The walls of the diaphysis are made up of dense, hard compact bone.

The wider section at the ends of the bone is called the epiphysis. It is filled with a type of osseous tissue called spongy bone. In between the diaphysis and epiphysis is the metaphysis. This contains growth plates, also known as epiphyseal plates, made up of hyaline cartilage.

The most superficial part of the bone is called the periosteum. This is a soft covering that provides blood flow to the bone, which is crucial in allow the bone to heal during injuries and fight infection. This layer is thick but eventually gets thinner as we age.

The next layer is hard and thick, and it is called cortical bone. It is similar to the hard shell of a turtle. This functions by holding up the muscles attached and it protects the body parts underneath it.

Cancellous bone comes after, and it is not as hard as the cortical bone. This has a spongy type of bone inside which allows faster healing during injury.

The deepest layer of the bone is the bone marrow. This is mainly responsible for production of all types of blood cells – red blood cells, white blood cells, and platelets.

Compact versus Spongy Bone

Most bones contain compact and spongy tissues but their concentration and distribution vary. The main difference between the two types is that the compact bone is stronger and denser as it is the outer layer whereas spongy bone is more porous being the inner layer of the bone.

A **compact bone**, also known as cortical bone, makes up the outer cortex of all bones. It is in immediate contact with the periosteum. A compact bone has a highly organized circular arrangement under a microscope. Each group f concentric circle makes up the structural unit of the compact bone called the osteon or the Haversian system. Inside the osteons are smaller rings made up of calcified matrix and collagen. These are termed lamella. Adjacent lamellae run at a perpendicular angle to each other, therefore allowing osteons to resist rotation forces in multiple directions. Found in the centermost part of an osteon is the central canal, or the Haversian canal, which allows blood vessels, nerves, and lymphatic vessels to pass through. These nerves and vessels branch off at right angles through another perforating canal called the Volksmann's canal. This canal extends to the periosteum and endosteum.

A **spongy bone**, also known as cancellous bone, contains osteocytes housed in a lacunae but are not arranged in concentric circles.

Instead, it is positioned in a lattice-like network of matrix spikes called trabeculae. These trabeculae are covered by endosteum which allows it to be easily remodeled. Despite its tangled appearance, these trabeculae form lines that direct out forces towards the compact bone the provide strength. Spongy bones also function by providing balance so that the bones can provide a smoother and easier movement. In addition, red bone marrow can be found in the spaces within a spongy bone and this is where hematopoiesis occurs.

Bone Classifications

There are 206 bones found in an adult human body. These bones are classified into five categories according to their shape and functions.

1.   Long bone. A long bone has a cylindrical shape and is longer than wider. It functions by

supporting the body's weight, and it also facilitates movement. These are usually found in the appendicular skeleton such as the upper limbs (e.g. humerus, ulna) and the lower limbs (e.g. femur, tibia).

2.   Short bone. This type is about as long as it is as wide. It acts by providing stability during motion. Examples are the carpal bones in the wrist, and the tarsal bones in the ankle.

3.   Flat bone. The term "flat" bone is a misnomer as although the bone is flat in shape, it is typically thin and curves as well. This type of bone serves as a point of attachment for muscles. They also protect the internal organs. An example is the sternum, the scapula, and the cranial bones.

4.   Irregular bone. Bones that cannot be easily characterized and do not fit any classification is called an irregular bone. These bones tend to have complex shapes such as the

vertebrae. They function by absorbing shock and compressive forces.

5.   Sesamoid bone. Bones with a small, round appearance, usually shaped like a sesame seed fall into this classification. These are embedded in tendons and they aide by absorbing compressive forces. An example would be the patella.

Bone Markings

The features of a bone's surface vary depending on its function and location in the body. There are three general classifications for this: The articulations, the projections, and the holes.

An **articulation** is where the surfaces of two bones meet. This forms a joint. To facilitate the function of articulation, one bone end is usually convex-shaped and the other is concave-shaped. A **projection** is an area where a bone projects is situated above another bone. This is where tendons and ligaments attach. The size of a projection depends on the force that it can exert. A **hole** is a groove or an opening that allows blood vessels and nerves to pass through and enter the bone. Just like projections, its size varies.

Listed below are more bone markings:

MARKING	DESCRIPTION	EXAMPLE
Crest	A ridge	Iliac crest
Condyle	A rounded surface	Femoral condy
Facet	A flat surface	Vertebrae
Fossa	An elongated basin	Mandibular fos
Fovea	A small pit	Fovea capitis o femur
Holes	Holes and depressions	Foramen

	A slightly elongated ridge	Temporal lines of parietal bones
cess	A prominent feature	Transverse process of the vertebra
is	Air-filled space in the bone	Frontal sinus
ie	A sharp process	Ischial spine
ercle	A small, rounded process	Tubercle of the humerus
erosity	A rough surface	Ischial tuberosity

<u>Blood and Nerve Supply</u>

The blood and nerve supply to the bones come from peripheral nerves. According to Hilton's Law, the nerve that supplies a muscle will also supply the underlying bone. Thus, is a group of muscles over a bone receive supply from a

119

specific nerve, the bone over which the muscles lie will also be innervated by the same bone.

Bone Formation and Growth

The skeleton is formed by bones and cartilages – the two of the strongest supportive tissues in the body. The skeletal system in embryos are made of hyaline cartilage, but eventually hardens into cartilage when they become young children. As adulthood is reached, the skeletal system becomes made of bones. However, the joints, ears, parts of the ribs, and the bridge of the nose remain made of cartilage.

The hardening or formation of bones is called **ossification**. This involves two major processes. First, the hyaline cartilage is covered in bone matrix by bone-building cells called osteoblasts. After birth, these hyaline cartilages are converted to bone except for two regions: the articular cartilage (covers of the bone ends) and the epiphyseal plates. This allows an increase in length as the infant grows as cartilage continuously form on these areas.

Bones also widen as they lengthen to maintain proportion. This occurs by adding bone matrix to the outside of the diaphysis from osteoblasts. The osteoclasts then remove bone from the inner face of the diaphysis wall, therefore expanding the medullary cavity. Both processes happen at the same time hence it is termed *appositional growth*.

Bone Remodeling

Despite being strong and sturdy, bones need to recover and remodel every once in a while. **Bone remodeling** is necessary to retain normal strength and proportion of nutrients to accommodate the body as it grows or changes. This process occurs for two reasons: 1) if there is a change in the calcium ion level in the blood; and 2) if there is a change in the pull of gravity and muscles on the skeleton.

When the calcium levels in the blood decrease below normal, the parathyroid glands are stimulated to release parathyroid hormone (PTH). PTH then activates osteoclasts which are responsible for breaking down bone matrix in order to release ore calcium ions into the blood.

On the other hand, when the body increases in size and weight, changes occur in the pull of gravity in the muscles and skeleton. The bones then remodel to become thicker and larger projections are made to increase strength. Osteoblasts then create new bone matrix to accommodate the changes.

Effects of Exercise on the Skeletal System

During exercise, mechanical stresses are received by the skeletal system. This helps improve the skeletal system as it strengthens it. When such stress is applied, collagenous fibers are produced by the body and mineral salts are deposited. As a result, the density and size of the bone increase. The higher the loads put on the skeletal system, the stronger it becomes. However, proper care and precaution must be observed on children as their bones are still developing. They should not be subjected to extra strenuous activities because these may provide negative impact on bone growth.

Body Levers

The muscles, bones, and joints of the body act together to form a lever and produce movement. A lever is composed of three parts: a fixed, rigid rod (usually the length of a bone) attached to a fulcrum (pivot) and a joint. Levers produce a mechanical advantage which magnifies movements. This means a small force can provide a much bigger force. There are three types of levers in the body:

1.   First Class Lever. In a class 1 lever, the axis (fulcrum) is located between the weight (resistance) and the force. This is the rarest type of body lever. An example is the atlantooccipital joint. When nodding the head, the axis is the joint, the weight is the head, and the force comes from posterior muscles such as the trapezius.

2.   Second Class Lever. For class 2 levers, the weight is found between the axis and the force. An example is the ankle joint that is in tiptoeing position. The weight in this position is the body, the ball of the foot acting as the fulcrum, and the force comes from the contraction of the gastrocnemius muscle.

3.   Third Class Lever. This is the most common type of lever. In this type, the force is applied between the weight and the axis. For

example, when performing a biceps curl, the fulcrum is the elbow joint with the forearm, wrist and hand as the weight and the biceps brachii as the force.

The muscular system and skeletal system work together to form the musculoskeletal system. This is the main framework of the human body. Without either one of the systems, movement is not possible. The musculoskeletal system however does not solely rely on muscles and bones. Other elements include the joints, tendons, cartilages, ligaments, and nerves.

Chapter 5. The Central Nervous System

The central nervous system or CNS is composed of two major parts: the brain and the spinal cord. This system is important for the control of bodily functions such as movement, thinking, sensing, speaking and memory. Its function is crucial for the survival of a living organism.

The brain is composed of four major regions. These are the cerebrum, the brain stem, the diencephalon and the cerebellum. The regulation of homeostasis and the activities controlling consciousness are governed by the brain.

The spinal cord is connected to the brain through the brain stem and is covered by the vertebrae of the spine. This part is responsible for the coordination of reflexes based on the sensory pathways that it contains.

Aside from the brain and spinal cord, the CNS also has protective structures that help protect its parts from injuries such as internal damage and physical trauma. This may be caused by accidents, pathogens or toxins that may travel the blood stream. These protective structures include the cranial bones and vertebrae.

The central nervous system is surrounded by a clear fluid called the cerebrospinal fluid. It is composed essentially of water and electrolytes where oxygen and carbon dioxide are dissolved as they are in the blood. The main purpose of cerebrospinal fluid is to remove metabolic wastes from interstitial fluids of nervous tissues and returning these wastes into the blood stream.

The two major parts of the central nervous system is divided into various parts. Each has their own important role in maintaining the healthy and proper functioning of the entire system.

The Brain

The brain is composed of the cerebrum, the brain stem, the diencephalon and the cerebellum. It is the overall control center of the human body. It is responsible for controlling a person's thoughts, memory and speech. It also controls the movement of the arms and legs, and the function of many organs within our body.

The Cerebrum

Often referred to as the gray mantle of the human brain, the cerebrum is the principal and largest part of the human brain and is responsible for the coordination of complex sensory and neural functions.

The cerebral cortex

The cerebrum is covered by a region of wrinkled gray matter called the cerebral cortex, also called the cerebral mantle, which is responsible for the more important functions and higher though processes of the nervous system.

The cerebrum is divided into two hemispheres by the longitudinal fissure. The cerebrum is also characterized by four lobes that contain cortical association areas, where information is collated and processed. These are the frontal lobe, parietal lobe, temporal lobe, occipital lobe.

The frontal lobe is found beneath the frontal bone of the calvaria, anterior to the cerebrum. The parietal love is found below the same bone, and sits between the frontal and occipital lobes. Beneath the temporal bone of the calvaria lies the temporal lobe, while the occipital lobe is located at the most posterior part of the cerebrum.

Acting as the sorting engine in the CNS, these lobes separate all the sensory perceptions the body receives and classifies them accordingly (e.g. personality vs. social response).

The cerebrum also has insular complex and limbic system. The insula's role is with regard to subjective emotional experience and body representation. This region maps out visceral states that are connected to emotional experience. This gives rise to a person having conscious feelings.

The limbic system is the part of the cerebrum that has three key functions. These functions govern emotions, memories and stimulation. This system is composed of parts that can be found within the cerebrum and just above the brain stem which connects parts of the brain that deal with high and low functions.

Within the cerebrum is a white matter called the corpus callosum. This serves as the principal pathway for communication between the right and left hemisphere of the cerebral cortex. The wrinkles of the cerebral cortex is referred to as gyri and the groove between two gyri is called a sulcus.

The frontal lobe

This is the part of the cerebrum that is in charge of cognitive skills which include problem solving, memory, language, emotional expression, judgment and behavior. It acts as the control panel of a person's ability to communicate and his entire personality.

It also controls a person's attention and concentration, self-monitoring, organization, and awareness of abilities and limitations. In connection with emotions, it is responsible for the functioning of inhibition of behavior, anticipation and judgment.

The parietal lobe

This is located at the back of the brain. This part is divided into two hemispheres that process sensory information from different parts of the body. It also interprets visual information and process mathematics and language.

In addition, it controls a person's sense of touch, spatial perception and the differentiation or identification of size, shapes, and colors. Visual perception is also governed by this lobe.

The temporal lobe

The temporal lobe is located at the back of the ears and extends to both sides of the brain. This part is in charge of sensory input, language, memory, comprehension, vision and emotion.

It is the part of the brain that helps in the understanding language and sequencing. It also shares the processing of hearing and organization with the other lobes.

The occipital lobe

The occipital lobe primarily processes vision. Therefore, any damage to even at least one side of this part may result to homonymous loss of vision. Any disorder of the occipital lobe can also result in illusions and hallucinations.

The subcortical nuclei

The subcortical nuclei lie beneath the cerebral cortex. These nuclei facilitate the overall activity of the cortex, which includes long-term memory, emotional responses, and focus on sensory stimuli. These are the basal nuclei, the basal forebrain and the limbic cortex.

Included in this region are the basal nuclei, which control impulses. These nuclei keep the body still while a person is listening to a lecture, for example. They help a person give greater attention to things and people, such as remembering what a person is saying.

The basal nuclei controls cognitive processing or planning movements and acts. The basal forebrain contains the nuclei that helps it process memory and learning. The limbic system is composed of structures, including the limbic cortex, that control emotion, behavior and memory.

The basal nuclei also host two streams of information processing. The direct pathway relates information from the striatum to the globus pallidus internal segment, while the indirect pathway carries information from the striatum to the globus pallidus external segment. The direct pathway causes the thalamus to excite the cortex through the disinhibition of the thalamus (part of the Diencephalon). When the information passes through the indirect pathway, the thalamus does not affect the cortex at all.

Sometimes a switch between the direct and indirect pathways happens. This is referred to as "substantia nigra pars compacta", and is responsible for the release of neurotransmitter dopamine. When the switch happens in the nuclei, the body will be in an active state. Hence there is more movement. Limited or zero movement means the substantia nigra pars compacta is silent.

The Brain Stem

The brain stem is composed of the midbrain, pons and the medulla oblongata. These parts collectively connect the brain to the spinal cord.

The midbrain is responsible for the coordination of the visual, auditory and somatosensory perceptual spaces to arrive at a sensory representation. It is the most superior part of the brainstem and is found at the posterior of the hypothalamus, which is part of the diencephalon (to be discussed in the next section).

This part of the brain stem is composed of reflex centers that control body movements. These movements are a result of the reaction produced from auditory and visual stimuli that are felt from the reflex centers in the eyes and head.

The word *pons* literally means bridge. This is because the pons is a bridge-like white matter connecting to the tegmentum region. The tegmentum region is a gray matter that contains neurons that receive input from the cerebrum and thalamus. The input is later sent to the cerebellum.

The last part of the midbrain is the medulla oblongata. The pons work with this part of the brain to regulate respiratory activities.

The medulla is the most inferior part of the brain that connects the latter to the spinal cord. It consists of sensory tracts that help in the entering and exiting of motor responses to the brain. It contains three integration centers that are vital for homeostasis called the respiratory center, the vasomotor center and the cardiac control center.

The respiratory center controls reflexes and the rhythm of breathing such as sneezing and coughing. The cardiac control center regulates the force and rate of hear contractions. The vasomotor center regulates blood pressure by the vasodilation of blood vessels and vasoconstriction of blood vessels and.

The reticular formation is another area that connects the medulla to the thalamus. It is responsible for regulating attention and general brain activity like wakefulness and sleep.

The Diencephalon

The word diencephalon translates into "through brain." It is a part of the adult brain that had its name from embryonic development.

The diencephalon connects the cerebrum and the rest of the nervous system. The different parts of the brain and the spinal cord all send information to the cerebrum through the diencephalon. However, there is an exception to this connection. The system associated with olfaction or the sense of smell is directly connected to the cerebrum and does not have to go through the diencephalon.

The diencephalon is divided into two regions: the thalamus and the hypothalamus. It also includes the subthalamus, which is a part of the basal nuclei, and the epithalamus, the host of the pineal gland.

The cone-shaped hypothalamus lies just below the thalamus, which makes up a part of the cerebral ventricle. The hypothalamus descends from the brain and ends in the pituitary stalk. The posterior part of the hypothalamus is called the median eminence, and contains neurosecretory cells. Adjacent to the median eminence are the third ventricle, the mammillary bodies, and optic chiasm.

The thalamus, a small structure located above the brain stem, is composed of ventricles and nerve fibers that project towards the cerebral cortex. The thalamus, unlike the hypothalamus, is bulb-shaped and measures around 5.7 cm in length. The thalamus is defined as a collection of nuclei that pass on information between the spinal cord, periphery, cerebral cortex and the brain stem. All sensory information pass through the thalamus in order to arrive at the cortex for processing with the exception of the sense of smell.

A synapse from the axons of the peripheral sensory organs is a requisite in order for sensory information to be processed. Again, information from the sense of smell is exempted.

The thalamus also processes information and the cerebrum also sends information to the thalamus. This are motor commands to be performed as a response to stimuli.

The hypothalamus is inferior and slightly anterior to the thalamus. Just like the thalamus, it is a collection of nuclei that are extremely involved in regulating homeostasis. This is the region of the diencephalon that controls the autonomic nervous system and the endocrine system. It is in charge of regulating the anterior pituitary gland, memory and emotion.

The Cerebellum

This part of the central nervous system is also considered as the "little brain" primarily because it is composed of gyri and sulci making it look like a miniature version of the cerebrum.

It has two hemispheres connected by a midline called the vermis, and is likewise composed of white and gray matter. While the gray matter folds over the cerebellum's surface, the white matter is located underneath its cortex, where the dentate, globose, emboliform, and fastigi nuclei – all the cerebellar nuclei – are also embedded.

The cerebellum is distinguished by 3 lobes, namely the posterior lobe, anterior lobe, and flocculonodular lobe. These are separated by the primary and posterolateral fissures.

On the side of the cerebellar vermis is the intermediate zone, and lateral to it are the lateral hemispheres. These hemispheres form the cerebrocerebellum, the biggest functional division of the cerebellum. The vermis and intermediate zone, on the other hand, compose the spinocerebellum, which receives proprioceptive information. The part of the cerebellum that is found in the flocculonodular lobe is called the vestibulocerebellum, which controls balance and reflex.

The cerebellum is responsible for the coordination of interactions of skeletal muscles or those connected to body movements. Neurons in the pons receive information from the cerebrum and these neurons project those information to the cerebellum. This provides motor commands from the cerebellum to the spinal cord.

Sensory information that enters the spinal or cranial nerves are copied to the inferior olive. Fibers from the inferior olive enter the cerebellum and the information is processed with that from the cerebrum.

If the frontal lobe controlling the primary motor cortex sends information of a command to the spinal cord to initiate movement, that command is sent to the cerebellum. The proprioceptive information for movement, the need for balance and the sensory feedback from joints and muscles are all sent to the cerebellum for processing in order to have a coordinated movement.

In cases when movement is uncoordinated because of externa factors, the cerebellum sends a corrective command to fix the difference between the sensory feedback and the original cortical command. It can be perceived from this that any damage to the cerebellum may result in problems with muscle tone, muscle contractions, movement coordination and balance.

The Spinal Cord

The spinal cord is connected to the brain through the brain stem. It is composed of nerves that connect to different parts of the body. It is the main path for carrying signals and between the brain and the nerves located at different parts of the body.

Anatomy Of The spinal cord

The spinal cord extends from the foramen magnum to the medulla, until the lumbar vertebrae. It is around 40 cm long, and 1 cm in diameter. At its sides arc two rows of nerve roots that join 31 pairs of spinal nerves. Cylindrical and composed of gray and white matter, the spinal cord has four regions, namely sacral, thoracic, lumbar, and cervical. Each region is segmented and contains the spinal nerves that dictate the movement and sensory perceptions of the body. These fibers carry sensations from the body to the brain, so that the brain can interpret these signals.

Motor neurons, in contrast, have projected axons into the periphery so that muscles can facilitate reflexes. The same neurons have descending axons that facilitate control of visceral regions in the body.

Since the spinal cord is segmentally organized, its distinct regions can be further identified. There are 12 thoracic regions, 5 sacral, 8 cervical, and one coccygeal nerve. The vertebral column also has ventral and dorsal roots that course through the intervertebral foramen.

Meanwhile, three meninges sheath the cord: dura, pia, and arachnoid. The dura is the outer sheath, while the pia is what covers the spinal cord's surface. In between lies the arachnoid. Lateral denticulate ligaments from the pial folds attach the dura to the spinal cord.

Spinal nerves exit below the vertebrae through 7 cervical segments. Only the eighth cervical nerve exists above the vertebrae. Root filaments of the spinal cord reach the intervertebral foramen, up until where the spinal nerves begin.

Peripheral nerve fibers coming from the dorsal root ganglia compose the dermatome. If a nerve is cut, the dermatome, or an area therein, loses sensation. Dermatomes are mapped on the body's surface, knowledge of which is used to determine which part of the spinal cord is damaged. Since neighboring dermatomes may overlap, it is possible that a larger part of the skin may lose sensation after an injury (e.g. the entire finger, and not just a portion of the finger).

The spinal cord also contains a central canal composed of cells. This is called the ependymal layer. This layer resembles the shape of a butterfly and is surrounded by gray matter. The "wings" of this butterfly shape are connected by the dorsal gray commissure and nerve fibers.

Neurons and glia are likewise found in this gray matter. They are further divided into four columns: the intermediate column, the ventral horn column, the dorsal horn, and the lateral horn.

The dorsal horn is found in all parts of the spinal cord; this part receives and processes somatosensory information. The lateral horn and intermediate column house the neurons that innervate pelvic and visceral organs. Lastly, the ventral horn contains the motor neurons in charge of skeletal muscles.

Functions of the spinal cord

The spinal cord has two basic functions. First, it carries nerve impulses from different parts of the body to the brain and vice versa. It connects a large part of the peripheral nervous system to the brain.

Second, it is responsible for serving as a reflex center for spinal reflexes. In this process, nerve impulses that reach the spinal cord through sensory neurons are transmitted to the brain. Signals that come from the motor areas of the brain travel back down to the spinal cord and leave the motor neurons.

The spinal cord also acts as a minor coordinating center that is responsible for simple reflexes like the withdrawal reflex. This type of reflex initiate nerve impulses in sensory neurons that later leads to the receptors in the spinal cord.

The impulses move through the spinal cord and the sensory nerve terminals synapse with the interneurons. These synapses either lead to flexors or suppress any motor output to extensors. Primarily, the withdrawal reflex intends to protect the body from damaging stimuli.

Aside from these functions, the spinal cord is also responsible for electrical communication and walking. In the latter, a collection of muscle groups in the legs are constantly contracting. The act of walking by taking step after step may seem incredibly simple but different factors have to be coordinated properly to allow this motion to occur. Central pattern generators are found in the spinal cord that are made up of neurons which send signals to the muscles in the legs, making them extend or contract, and produce the alternating movements which occur when a person walks.

Chapter 6. Peripheral Nervous System

The peripheral nervous system (PNS) is the portion of the nervous system that does not belong to the central nervous system (CNS). That is, it comprises the nerves and ganglia that are not located in the spinal cord and brain. The PNS is mostly outside the vertebral and cranial cavities, but some parts are within them as well.

Unlike the CNS, the PNS is not encased by the skull and vertebral column. It does not have a defensive layer like the blood-brain barrier. Because of this, it is more vulnerable to injuries and impurities.

The PNS have plexuses or nerve fibers that supply the bones, joints, muscles, and skin. The thoracic region doesn't have a plexus.

Functions of the PNS

The nervous system as a whole is responsible for sensation, response (or motor function), and integration, which is the combination of sensations to higher mental functions such as learning, memories, and emotions.

The CNS and the PNS has the same functions, but their components are found in different regions in the body. In particular, the PNS connects the CNS to the bodily organs and limbs.

The PNS may be classified according to function: the somatic nervous system (SNS) and the autonomic nervous system (ANS). The SNS enables conscious actions while the ANS manages automatic processes.

Structures

As mentioned, the PNS is composed of nerves and ganglia:

Ganglia

Ganglia are groups of neurons in the PNS. They may be categorized as sensory or autonomic. The counterparts of the ganglia in the CNS are the nuclei.

Posterior or dorsal root ganglia are the most common sensory ganglia. These neurons have axons with sensory endings in the peripheral body parts (e.g. the skin) and have dorsal nerve roots that reach thc CNS.

Nerves

Nerves are bundles of axons within the PNS. The counterparts of nerves in the CNS are called tracts. The nerves are composed of nervous tissues, connective tissues, and blood vessels.

Cranial Nerves

Cranial nerves are PNS nerves that are linked to the brain. They are mainly responsible for the motor and sensory functions of the head and neck, with one targeting organs in the abdomen and thorax.

There are 12 Cranial Nerves:

I (Olfactory)

II (Optic)

III (Oculomotor)

IV (Trochlear)

V (Trigeminal)

VI (Abducens)

VII (Facial)

VIII (Auditory/Vestibulocochlear)

IX (Glossopharyngeal)

X (Vagus)

XI (Spinal Accessory)

XII (Hypoglossal)

Cranial nerves I (smell), II (vision), and VIII (balance and hearing) have sensory functions. Nerves III, IV, VI (eye movements), XI (head and neck movements), and XII (lower throat movements) are motor nerves. The rest have both sensory and motor functions: nerves V (facial movement and sensation), VII (facial movement, taste), IX (throat movement, taste), and X (autonomic motion/sensation of the viscera).

The cranial nerves that go beyond the brain are treated as part of the PNS. The 2nd cranial nerve (the optic nerve) is not included in the PNS.

Spinal Nerves

Spinal nerves originate from the spinal cord and carry somatosensory information. All of them have motor and sensory axons that branch into two nerve roots: the dorsal and ventral nerve roots.

The dorsal nerve root is made up of sensory axons that go into the spinal cord. The ventral nerve root is made up of motor fibers.

The spinal nerves have 31 pairs. Cervical nerves consist of 8 pairs and named C1 to C8. 12 pairs of thoracic nerves are labeled T1 to T12. 5 pairs of lumbar nerves are called S1 to S5. There is one pair of coccygeal nerves as well.

In the cervical area, the nerves are named after the vertebrae near them. The spinal nerve roots get their name from the vertebrae below them.

As such, the Spinal Nerve C1 is located between the 1st cervical vertebrae and the skull, C2 is between the 1st and 2nd cervical vertebrae, and so on. However, C8 comes out of the space between the 7th cervical vertebra and the 1st thoracic vertebra.

In the thoracic to coccygeal area, the spinal nerve roots appear below the vertebrae they are named after, while in the lumbar and sacral area, they run through the dural sac. Below L2, the spinal nerves are called cauda equina.

Cervical Spine Nerves (C1 – C4)

The cervical spine nerves are found in the back of the head and the neck. The first spinal nerve C1 is also called suboccipital nerve. They enable movement of the muscles found at the bottom of the skull.

Nerves C2 and C3 Form the neck nerves – these provide both motor and sensory control. It includes the greater occipital nerve, which is responsible for providing sensation to the back of the head, and the lesser occipital nerve, which gives sensation to the region behind the ears, and the greater and lesser auricular nerves related to the ear.

The phrenic nerve is important for survival because it enables breathing. It starts from nerve roots C3, C4, C5 and supplies the thoracic diaphragm. Take note that if the spinal cord is cut above C3, it becomes impossible to breathe spontaneously.

Brachial Plexus (C5 – T1)

Also known as plexus brachialis, the brachial plexus is composed of nerves C5 to C8 as well as T1, the first thoracic spinal nerve. It controls the upper back and upper limbs.

Lumbosacral plexus (L1 – Co1)

The lumbosacral plexus has nerves L1 to Co1. It is comprised of the anterior parts of lumbar nerves, sacral nerves, and coccygeal nerve. The first lumbar nerve is joined by a nerve branch from 12th thoracic nerve (T12).

Divisions of the Peripheral Nervous System

As mentioned, the peripheral nervous system has two systems – the somatic nervous system and the autonomic nervous system. Both of these systems have sensory/afferent and motor/efferent parts.

Sensory/Afferent - The sensory or afferent division of the PNS receives impulses from receptors and sends them to the CNS. This involves visceral and somatic sensory nerve fibers.

Motor/Efferent - The motor or efferent division of the PNS accepts impulses from the CNS and transmits them to effectors, which are the muscles and glands.

Aside from being divided into afferent and efferent parts, the PNS has somatic and autonomic components:

Somatic Nervous System

The somatic nervous system (SNS) enables the voluntary and conscious control of the skeletal muscles. It sends brain signals to the muscles and other end organs.

Although the SNS's main purpose is to carry out movements, it has nerves with a somatic function, sensory function, and both functions (somatosensory).

Autonomic Nervous System

The autonomic nervous system (ANS) is also known as visceral system since it mainly relates to the viscera. It exerts automatic (involuntary) control over glands and smooth muscles involved with vital physiological functions such as heartbeat, salivation, digestion, and urination. It is also responsible for pupil dilation and constriction.

In the ANS, ganglionic neurons connect the brain and spinal cord to organs such as the cardiac, exocrine, and endocrine organs. Impulses are conducted from the CNS into the muscles of the heart, the smooth (not skeletal) muscles, and the glands.

Functional States

The Autonomic Nervous System is always activated, but it functions in either sympathetic or parasympathetic condition. The neurotransmitters released depend on the state of the ANS.

Sympathetic

The sympathetic division of the ANS activates during the fight or flight state that is provoked by the perception of stress or danger. It mobilizes the systems of the body to make it respond to the situation by fighting the threat or escaping from it.

The SNS increases blood flow and heart rate and decreases the bodily activities not essential for survival, such as digestion.

The rise of heart rate is accompanied by the dilation of the bronchia, which increases air flow in and out of the lungs. These maximize the amount of energy that can be used to ensure survival. The pupils dilate to increase visibility.

Aside from inhibiting the digestive organs and the pancreas and gallbladder, it stops saliva production and relaxes the urinary bladder. This is done to direct energy away from maintenance work into making quick and forceful actions.

The PNS' sympathetic division is also referred to as the thoracocolumbar division because the nerves involved originate in the thoracal and lumbar region of the spinal cord.

Parasympathetic

The ANS' parasympathetic division is geared towards conserving energy and maintaining the body's health. It is activated during rest and digestion.

When the body is in parasympathetic mode, heart rate, respiration rate, and other sympathctic response activities decrease. Activities relating to digestion and salivation increase. Saliva production returns, gall bladder and pancreas activity resume, and the urinary bladder constricts. Dilated pupils go back to normal size.

Unlike the sympathetic nervous system, the parasympathetic division can be controlled consciously, such as during urination and defecation. Thus, one can counteract the sympathetic response by deliberately relaxing the muscles and breathing deeply.

The PNS' parasympathetic division is also known as the craniosacral division since its nerve fibers involve a few cranial nerves (III, VII, IX, and X) and sacral nerves (S2 to S4)

Enteric Nervous System (ENS)

The Enteric Nervous System is a subdivision of the Peripheral Nervous System. It is composed of peripheral nerves and ganglia that are embedded within the gastrointestinal lining:

Myenteric Plexus

Sandwiched between the layers of the muscularis externa, the myenteric plexus increases the tone of the gut and controls the intensity and velocity of the digestive tract contractions.

Submucosal Plexus

The submucosal plexus is in the submucosal layer of the digestive tract. It enables nutrient absorption and waste secretion. It also controls some muscle movements.

The Enteric Nervous System enables local control of the digestive system without requiring input from the parasympathetic and sympathetic branches of the Peripheral Nervous System. It can function independently of the spinal cord and brain, but despite this, it can respond to and receive signals from the rest of the body. Because the PNS stimulates the ENS, defecation can still be done even during fight or flight mode when the SNS is inhibiting digestive function.

In summary, the Peripheral Nervous System is the part of the nervous system that is not included in the Central Nervous System (the brain and spinal cord). It is composed of the Somatic (voluntary) and autonomic (involuntary) nerves that have sensory, motor, or sensorimotor functions, the Ganglia, groups of neurons that receive sensory stimuli and relay motor responses to the muscles and viscera, and the Enteric Nervous System, which is the part of the PNS that is contained within the gastrointestinal tract and can function separately from the rest of the nervous system

Since all the parts of the nervous system are connected to one another, the PNS have parts that extend to the CNS. It can be quite tricky to classify nerves as central or peripheral, but most structures have distinct functions that have allowed them to be named accordingly.

Chapter 7. The Autonomic Nervous System

Formerly known as the vegetative nervous system, the autonomic nervous system (ANS) is a division of the peripheral nervous system (PNS). The ANS influences the involuntary functions of internal organs.

This division of the peripheral nervous system regulates body functions such as sexual arousal, urination, pupillary response, respiratory rate, digestion, and heart rate. It's also responsible for the human body's fight-or-flight response.

Within the brain, the hypothalamus regulates the ANS. Cardiac regulation, vasomotor activity, and control of respiration are examples of autonomic functions. Reflex actions, such as vomiting, swallowing, sneezing, and coughing, are also autonomic functions.

These functions are subdivided into areas and are linked to the subsystems of the autonomic nervous system. The hypothalamus, which is located above the medulla oblongata, integrates autonomic functions. To do so, it receives regulatory inputs from the limbic system.

The Three branches of The Autonomic Nervous System

The ANS has three branches.

1. The enteric nervous system (ENS)
2. The parasympathetic nervous system (PSNS)
3. The sympathetic nervous system (SNS)

The sympathetic nervous system is the fight-or-flight system.

Whereas, the PSNS is considered the feed-and-breed and the rest-and-digest system.

In most cases, the first two branches contradict each other. When one system activates an automatic reaction (physiological response), the other will inhibit it. However, there are exceptions. In orgasm and arousal for example, both systems play a role.

In general terms, the SNS is a quick response mobilizing system. The PSNS, on the contrary, is a slowly activated, dampening system.

The ENS is the third branch of the autonomic nervous system. It's also a subdivision of the ANS and is called the 2nd brain. It controls the functions of the digestive tract, and it can act independently. It doesn't rely on the other divisions, namely SNS and PSNS. In spite of that fact, the other two systems still influence the ENS.

Generally, the ANS is the region of the nervous system that maintains and controls involuntary functions. Even though the ANS is also called the visceral nervous system, it's also linked with the body's motor side. Involuntary functions are autonomous; however, some autonomous body functions often work together with the voluntary nervous system.

Key Functions of the Three Branches

Typically, the parasympathetic and sympathetic divisions oppose each other. This opposition, however, is complementary in nature, not antagonistic. The SNS serves as the accelerator, and the PSNS functions like a brake.

The SNS functions with actions that require quick responses, whereas the PSNS functions in actions not requiring immediate reaction. The ENS controls the digestive system directly.

Involuntary controls are largely attributed to the autonomic nervous system; however, many instances of parasympathetic and sympathetic activity can't be ascribed to rest or fight situations.

As an example, standing up would require an unsustainable drop in blood pressure levels if it weren't for the compensatory increase in the arterial sympathetic tonus. Another example is the second-to-second modulation of heart rate by parasympathetic and sympathetic influences.

Generally, to achieve homeostasis, the SNS and PSNS permanently modulate vital functions, in an antagonistic fashion. Higher organisms, such as humans and elephants, maintain integrity through homeostasis.

Homeostasis, in biology, is the state of steady internal chemical and physical conditions maintained by self-organizing life forms, namely mammals, reptiles, amphibians, etc. Homeostasis relies on balancing feedback regulation. This, in turn, depends on the ANS.

The Sympathetic Nervous System

Alongside the ENS and PSNS, the SNS helps regulate the body's internals. The SNS has connections with the pelvic plexuses, abdominal aortic, and thoracic, and it extends from the thoracic to the lumbar vertebrae.

Stress is the physiological instinctive response to a threatening situation, such as circumstances that activates the fight-or-flight system. The PSNS counteracts the SNS's fight-or-flight responses.

Here are the other primary functions of the SNS.

- The nerves from the SNS innervate the tissues in every organ system in the body.
- The noradrenaline and adrenaline hormones secreted by the adrenal medulla mobilize the body and facilitate physical activity in response to threatening situations.
- The SNS helps maintain homeostasis.
- It also regulates autonomous actions.

One of the major roles of the SNS is to regulate the numerous homeostatic mechanisms in living things. In every system, the SNS innervate tissues. As well, it provides physiological regulation over various body processes, such as gut motility, urinary output, pupil diameter, and muscle contractions.

Perhaps, the SNS is best known for mediating hormonal stress responses—fight or flight responses—and neuronal responses. Its activation occurs when the preganglionic sympathetic fibers, located at the end of the adrenal medulla, secrete acetylcholine (ACh). Acetylcholine is an organic chemical that acts as a neurotransmitter in the body and brain of humans and animals. A neurotransmitter is a chemical message released by neurons to send signals to other cells. ACh triggers the secretion of noradrenaline (norepinephrine) and adrenaline (epinephrine).

The fight-or-flight response, which is also called the acute stress response or hyperarousal, is a physiological response. It occurs in reaction to a threat to survival. First described by Walter Bradford Cannon, hyperarousal occurs when higher organisms react to harmful events with a general discharge of the SNS. This prepares living organisms for fleeing or fighting.

Specifically, the adrenal medulla produces a cascade of hormones. This triggers catecholamine-secreting cells to secrete epinephrine and norepinephrine.

Additionally, the neurotransmitters serotonin and dopamine and the hormones cortisol, testosterone, and estrogen also affect how an organism reacts to stress.

Consequently, the SNS brings about the fight or flight response via neurotransmission. The catelochamines, such as epinephrine and norepinephrine, secreted from the adrenal medulla, indirectly triggers hyperarousal. In turn, hyperarousal acts on the circulatory system.

Impulses travel through the sympathetic nervous system bidirectionally. Efferent messages, which refer to the signals that are carried away the central nervous system (CNS) as opposed to afferent messages, which are carried towards the CNS, trigger changes in parts of the body. As an example, the SNS can decrease motility, widen bronchial tubes, and increase heart rate. It can also cause perspiration, goosebumps (piloerection), and pupillary dilation.

The Parasympathetic Nervous System

The PSNS regulates gland and organ functions during rest. Its primary neurotransmitter is acetylcholine. Nonetheless, cholecystokinin and other peptides also act on this division of the autonomic nervous system.

Cholecystokinin (CCK) is a protein hormone—peptide hormone. Peptide hormones affect the endocrine system. CCK stimulates the digestion of protein and fat. It triggers the release of bile and digestive enzymes.

The key functions of the parasympathetic nervous system are as follows:

- The PSNS stimulates defecation, digestion, urination, lacrimation, salivation, and sexual arousal.
- The vagus nerve aids in controlling the lungs, heart, and organs for digestion.
- The PSNS regulates blood vessels, digestive muscles, adrenal glands, salivary glands, and heart muscles.
- During rest and digestion, the parasympathetic nervous system makes the pupils to narrow.

Collectively, the SNS and PSNS are known as the autonomic nervous system. The ANS regulates body functions unconsciously.

Being a division of the ANS, the PSNS works with the SNS to maintain homeostasis. For instance, during hyperarousal, the SNS increases blood pressure and accelerates heart rate.

Afterward, the PSNS decreases heart rate and lowers blood pressure. It also restarts bodily functions that the SNS put on hold. This includes digestion and salivation.

During rest periods, the body devotes energy to physiological processes not involved with fleeing or fighting.

PSNS nerves begin at the middle of the spinal column and in the brain medulla. The vagus nerve is located in the medulla. To reiterate, this nerve controls the digestive organs, lungs, and heart.

The spinal cord and the brain are important

parts of the nervous system. They compose the CNS or central nervous system. PSNS nerves that originate in the brain are referred to as cranial nerves.

Ganglia, or structures containing nerve cell bodies, which are linked by synapses, are extensions of PSNS nerves. Ganglia are located in or near body organs so that the target areas can receive signals from the parasympathetic nervous system.

As stated, the SNS and PSNS regulate many body parts, including glands and muscles. Specifically, the PSNS controls the adrenal glands and salivary glands during fight-or-flight situations. It inhibits or decreases motility and saliva production.

During digestion and periods of rest, it makes the pupils constrict, and during hyperarousal, the pupils dilate to improve vision and allow higher organisms to make swift decisions. This division of the ANS also causes increased food digestion, mucous production, and urine secretion.

The PSNS should not be confused with the PNS or the peripheral nervous system. The spinal cord and the brain compose the CNS. By contrast, the PNS contains all the parts of the nervous system, except the brain and spinal cord.

Therefore, the parasympathetic nervous system is part of the peripheral nervous system. PSNS nerves originate in the spinal cord or the brain, but most parts of the PSNS isn't located in the said regions. Additionally, the PNS includes the SNS and SoNS. In contrast to the function of the ANS, the SoNS regulates voluntary body movements.

In summary, the PSNS controls the "rest and digest actions" of the body. On the contrary, the SNS regulates fight-or-flight actions. Along with the ENS, the SNS and PSNS compose the autonomic nervous system.

The Enteric Nervous System

The enteric nervous system is a subdivision of the ANS. The ENS directly controls the digestive system. It can operate independently of the spinal cord and the brain.

For a handful of reasons, the ENS is referred to as the second brain. It can operate autonomously. In addition, according to vertebrae studies, even if the vagus nerve is severed, the ENS will continue to function.

The enteric nervous system is considered as the nervous system of the gastrointestinal tract. Being a part of the autonomic nervous system, it also regulates involuntary movements. It can coordinate reflexes.

In vertebrates, the ENS includes interneurons, afferent neurons, and efferent neurons. They enable the system to carry reflexes and act as an integrating center when CNS input is absent.

Moreover, the sensory nerve cells report on chemical and mechanical conditions. The motor neurons, through digestive muscles, control churning and peristalsis. Other parts regulate enzymes secretion.

The ENS also benefits from over 30 neurotransmitters. These neurotransmitters are identical to the chemical substances in the central nervous system, namely serotonin, dopamine, and acetylcholine.

Approximately 90% of the human body's serotonin can be found in the stomach. About 50% of the body's dopamine also lies in the gut.

Serotonin is a neurotransmitter that constricts blood vessels. It also influences an individual's motor skills, emotions, and sleep patterns. According to Healthline, it affects every part of a living organism.

Dopamine acts as a precursor of other substances, such as epinephrine. Dopamine or DA functions both as a neurotransmitter and a hormone. It plays several roles in the body and the brain.

Through afferent and efferent neurons, the enteric nervous system can alter body responses, namely, bulk and nutrient composition. Support cells also compose this division of the ANS.

These cells are similar to the brain's astroglia. Astroglia are glial tissues composed of astrocytes. Support cells are responsible for many functions, which include biochemical support.

The autonomic nervous system has motor, gland-stimulating, and sensory components. It's composed of three branches: the parasympathetic, sympathetic, and enteric nervous system. The SNS regulates quick responses and is also considered the fight-or-flight response system. The PSNS is described as the dampening system, for it regulates unconscious actions. Lastly, the ENS is responsible for governing the function of the digestive system.

Chapter 8. The Endocrine System

The endocrine system is composed of the glands located in the different parts of the body. This organ system affects various physiological processes and bodily functions, such as metabolism, temperature regulation, and reproduction.

The endocrine system is also described as a chemical messenger system that is composed of feedback loops of hormones. Hormones are regulatory substances in an organism. They're transported in tissue fluids, namely sap or blood. They stimulate specific tissues or cells into action. Glands directly secrete hormones into the bloodstream of an organism.

In humans, the thyroid and the adrenal glands are the major endocrine glands. In vertebrates, the hypothalamus, the region of the forebrain which coordinates the activity of the pituitary and the autonomic nervous system (ANS), is the neural control center for all endocrine organs.

This chapter covers how the endocrine system works and how the different glands operate.

Feedback Loops

Organ systems function on a mechanism of outputs and inputs. Feedback loops are any biological occurrence in which an output amplifies a system. This is called positive feedback. A negative feedback, on the contrary, inhibits an organ system.

Both negative and positive feedback loops are vital in living organisms. They help maintain homeostasis. They're utilized extensively in regulating the secretion of hormones.

Functions Of Hormones

Hormones act as messengers, affecting the activity of target sites. They activate, control, and coordinate functions throughout the body. Some hormones help bones grow, whereas others trigger the production of blood cells.

They keep organ systems running. They also relay signals to vital organs. These signals dictate how parts of the body should work. Hormones don't just appear magically.

Glands nurture and produce hormones, and they are also the ones responsible for sending the regulatory substances to the appropriate cells, tissues, or organs.

Structures of the Endocrine System

The endocrine system consists of tissues, cells, and organs that produce hormones. The endocrine glands, which include the hypothalamus, pancreas, testes, and ovaries, are the most important parts of this organ system.

Endocrine glands are ductless organs. They secrete hormones directly into the bloodstream, rather than through a duct. The blood vessels and the interstitial fluid then transport the chemical substances in every part of the body.

This organ system includes the pineal, adrenal, parathyroid, thyroid, and pituitary glands. Some endocrine glands have both non-endocrine and endocrine functions.

As an example, the pancreas is composed of cells that aid in digestion, as well as cells that produce glucagon and insulin. Glucagon is a hormone that promotes the breakdown of glycogen to glucose, whereas insulin regulates the amount of glucose in the blood.

There are also organs that contain endocrine cells. These include ovaries in females, testes in male, skin, liver, the small intestine, stomach, kidneys, heart , thymus, and hypothalamus.

Endocrine glands are entirely different from the glands that secrete substances onto the outer layer of the body through ducts. The latter is called exocrine glands, of which the skin's sweat and sebaceous glands are examples.

However, some glands and organs in the endocrine system also have an excretory function. The pancreas is an example. The acinar cells synthesize and secrete the enzymes through the accessory and pancreatic ducts to the small intestine's lumen.

How the Endocrine System Works

The primary purpose of this organ system is to maintain the body's chemical balance. To do so, it sends chemical substances called hormones in every part of the body through the bloodstream. From a distance, specific hormones control specific target sites.

For instance, insulin signals the liver and fat and muscle cells to absorb glucose from the blood. With this, blood glucose levels are stabilized.

Glands can be found in the different parts of the body. The pituitary gland is located in the brain, between the pineal gland and the hypothalamus. These glands are referred to as ductless organs since they don't have a duct system. They secrete hormones into the bloodstream directly.

Endocrine glands regulate various physiological processes. The pituitary gland, for example, stimulates growth, lactation, and water retention. There are two types of endocrine glands: peripheral and central.

The hypothalamus and the pituitary gland are the central glands; both are located in the brain. The latter is called the master gland, for it secretes various hormones that regulate the functions of other glands.

As an example, the adrenocorticotropic hormone secreted by the anterior pituitary gland stimulates the adrenal glands to produce steroid hormones, such as cortisol. Steroid or growth hormones regulate body composition, metabolism, and growth.

The pituitary, along with the hypothalamus, works together to maintain and regulate bodily functions, namely reproduction, metabolism, and salt and water balance.

On the contrary, the peripheral glands, which include the pineal glands, adrenals, thyroid, and parathyroid, have one function only—to produce hormones. Nevertheless, they may have primary or secondary functions as well.

The ovaries, for example, have two reproductive functions. Firstly, they produce egg cells (oocytes) for fertilization. Second, they secrete progesterone and estrogen.

The corpus luteum in the ovary secretes progesterone. This hormone plays a vital role in menstruation and in preventing preterm labor. It prepares the uterus for implantation by thickening the endometrium. Additionally, progesterone inhibits lactation. The decrease in progesterone levels after delivery triggers lactation.

Oxytocin, on the contrary, aids in milk production. The hypothalamus produces oxytocin, and the posterior pituitary releases this hormone. Oxytocin plays a role in childbirth, sexual reproduction, and social bonding.

In females, estrogen makes the bones shorter and smaller, the shoulders narrower, and the pelvis broader than males. It also promotes fat storage around the thighs and hips.

The Different Endocrine Glands and the Hormones They Produce

The Pituitary and the Hypothalamus

The master gland is consists of a posterior and an anterior lobe. As stated, it's located in the sella turcica below the hypothalamus. The pituitary gland controls other glands, whereas the hypothalamus stimulates the master gland to secrete other hormones.

This is the reason why they're called the central glands. They work together to regulate physiological processes through hormone secretion. The anterior lobe secretes seven hormones. Five of these are regulators. They induce other endocrine glands to produce hormones.

Here are the tropic hormones produced by the adrenal cortex:

a) The somatotropin—the growth hormone (hGH)--stimulates cell growth.

b) Thyrophine is also known as the thyroid-stimulating hormone (TSH). It stimulates the thyroid to secrete and produce triiodothyronine and thyroxine.

c) The Adrenocorticotropic hormone or ACTH triggers the adrenal cortex to secrete aldosterone and cortisol.

d) The follicular-stimulating hormone (FSH) is responsible for ovum growth in the ovaries of females. In males, FSH promotes spermatozoa formation. FSH is also classified as a gonadotropic hormone, for it influences hormone secretion and the growth of the testes and ovaries.

f) LH or luteinizing hormone triggers ovulation and causes the secretion of progesterone. In males, it stimulates testosterone secretion. Like FSH, LH is also considered a gonadotrophic hormone.

Two of the seven hormones don't stimulate endocrine glands to produce hormones. Therefore, they aren't tropic hormones. The non-tropic hormones are the prolactin (PRL) and the melanocyte-stimulating hormone (MSH).

Prolactin promotes breast tissue growth, and after childbirth, it sustains lactation. MSH causes skin pigmentation and influences melanin formation.

The posterior lobe of the pituitary gland secretes and stores the two hormones: the oxytocin and vasopressin. Vasopressin or antidiuretic hormone (ADH) prevents excessive water loss.

The Thyroid Gland

There are two pear-shaped lobes in the thyroid gland. The isthmus, a strip of tissue, separates the two lobes. The thyroid gland is located on the two sides of the trachea and below the thyroid cartilages.

The gland is consists of tiny sacs that are filled with colloid, a gelatin-like fluid. The hormones from the thyroid are store in the sacs until they're needed by the body.

The triiodothyronine (T3) and thyroxine (T4) are iodine-rich hormones. Both T3 and T4 are needed in maintaining normal metabolism. These two supports the body's BMR or basal metabolic rate. They also aid cells in oxygen uptake.

Injections of thyroxine increase the metabolic rate of an organism, and the removal of the thyroid gland decreases T3 and T4 levels in the body. This leads to heat loss, obesity, and poor mental and physical development.

In addition, the thyroid secretes calcitonin. This hormone aids the body in maintaining calcium balance.

The Parathyroid Glands

Four half-inch oval bodies compose the parathyroid glands. They are the tissues that secrete PTH or the parathyroid hormone which regulates calcium levels in the body.

The Pancreas

Found behind the stomach, the pancreas is also a part of the gastrointestinal system. As part of the endocrine system, this gland produces glucagon and insulin.

These two hormones regulate sugar and starch metabolism. Insulin decreases blood sugar levels. Contrastingly, glucagon increases blood sugar levels by stimulating the liver to synthesize glucose. This process is called gluconeogenesis.

Adrenals

Humans have two adrenal glands. Each adrenal is consists of two parts: the adrenal medulla and the adrenal cortex. The medulla and cortex secrete a different hormone from the other.

The medulla secretes catelochamines. Catelochamines are chemicals that act as modulators of hyperarousal or the fight-or-flight response.

The cortex produces mineral corticoids. These regulatory substances maintain the mineral potassium and sodium levels in the body. Aldosterone, a corticosteroid hormone, stimulates sodium absorption of the kidneys. Therefore, it regulates salt and water balance.

Cortisol, which is also secreted by the cortex, promotes sugar production from stored fat. The adrenal cortex also secretes androgens that promote the development of secondary male characteristics.

The adrenal medulla secretes adrenaline and norepinephrine (noradrenaline). These hormones work with the sympathetic nervous system (SNS). When an organism is under stress, the adrenal medulla secretes the aforementioned hormones. They aid the body in responding to stressful situations by increasing blood pressure, blood sugar level, and heart rate.

Gonads

Gonads are organs that produce and secretes gametes or reproductive cells. The female ovaries secrete egg cells; the male testes produce sperm cells. Other than that, gonads also release gonadotrophins or hormones that stimulate reproductive organs.

The ovaries secrete estrogen. Estrogen is needed in the development of secondary female sex characteristics, such as breast and pubic hair development. This hormone also regulates menstruation.

The corpus luteum, a hormone-secreting structure that develops in an ovary after an egg cell has been discharged, releases progesterone. The main function of this hormone is to thicken the endothelium of the uterus for pregnancy.

If fertilization does not occur, progesterone secretion stops. Then, menstruation follows. Progesterone and estrogen are present at low levels in the body of males.

Testosterone, a chemical substance secreted by the testes, is the primary male sex hormone. It's also an anabolic steroid. Testosterone increases bone and muscle mass and promotes the growth of body hair. Moreover, it promotes normal sperm development.

As progesterone and estrogen, the primary female sex hormones, low levels of testosterone are also present in females.

To summarize, the endocrine system is composed of ductless glands. Glands are organs that secrete hormones. Hormones serve as messengers and regulators. They're secreted directly into the bloodstream. Hormones stimulate specific cells, tissues, or organs into action. There are two types of glands in the body: central and peripheral. The hypothalamus and pituitary gland are the central endocrine glands. There are also two types of hormones secreted by the glands: non-steroidal and steroidal.

Chapter 9. The Cardiovascular System

The cardiovascular system is an organ system that facilitates the circulation of blood and nutrients throughout the body. This is also called the circulatory system or the vascular system.

The Components

The main components of the cardiovascular system are the blood, the heart, and the blood vessels. Each of these will be discussed in detail in their dedicated chapters.

The blood which consists of plasma, erythrocytes (RBC), leukocytes (WBC), and thrombocytes (platelets) is the circulating fluid that transports important gases, nutrients, and waste products to and from and the body. The heart is a muscular organ that collects blood from the body tissues and pumps it to the lungs for oxygenation.

The heart also pumps oxygenated oxygen-rich blood from the lungs to the systemic circulation and all parts of the body. The blood vessels – arteries, arterioles, veins, venules, capillaries are components of the vascular system that transport blood throughout the body.

The lymphatic system is also a part of the circulatory system. The lymph is a recycled excess plasma that has been filtered from the cells and brought back to the lymphatic system. The lymph is circulated by the vascular system to give the body extra immunity and to keep the bodily tissues healthy and functional.

Pulmonary Circulation Vs. Systemic Circulation

The blood circulatory system is composed of two components: the pulmonary circulation and the systemic circulation.

The pulmonary circulation or the circulatory system of the lungs is the process of transporting oxygen-depleted blood from the right ventricle to the lungs for oxygenation. The pulmonary artery carries the deoxygenated/unoxygenated blood from the right ventricle to the lungs while the pulmonary vein transports the oxygenated blood from the lungs to the left atrium. The process of true gas exchange – where carbon dioxide is released and oxygen is absorbed, happens in the lungs – specifically in the alveoli.

The systemic circulation is the process of transporting oxygenated blood from the heart to the different body tissues and collecting deoxygenated/unoxygenated blood from different tissues to the heart. The aorta in particular is responsible for transporting O_2-rich blood from the left ventricle where it is stored after pulmonary circulation to the body tissues. Meanwhile, the superior vena cava and inferior vena cava receive unoxygenated blood from upper and lower portions of the body.

Transportation Of Electrolytes, Hormones, Nutrients, And Wastes

Aside from transporting major gases (carbon dioxide and oxygen), the cardiovascular system also transports electrolytes, nutrients, and body wastes. The blood is made up of 55% plasma – a yellowish liquid component of the blood that contains water, salts, nutrients, hormones, and nitrogenous wastes. The salts in the plasma contain different electrolytes - sodium, potassium, magnesium, calcium, chloride, and bicarbonate. These electrolytes are essential in maintaining different body functions such as muscle contraction (calcium) and nerve transmission (potassium).

The cardiovascular system (CVS) is a network of blood vessels that link all organs of the human body. This internal network serves as a delivery system for different nutrients such as glucose and minerals essential to keeping the body healthy. Different chemical messengers, more commonly called hormones of the endocrine system are transported by the CVS to their respective organs. The vascular system also works together with the respiratory and urinary systems to expel carbon gases and waste products.

Protection Against Opportunistic Microorganisms And Blood Loss

Another important function of the cardiovascular system is providing immunity and protection. The blood cell also called hematocyte, hemocyte, or hematopoietic cell is produced by the hematopoiesis process. This is the formation of 3 main blood cell components – erythrocytes, leukocytes, and thrombocytes. The white blood cells or the leukocytes protect the body against foreign bodies and disease-carrying microorganisms. They are prompted immediately when there is a presence of infection, and they enable the immune system to act quickly.

Platelets or thrombocytes are body cells responsible for blood clotting and coating different body lacerations. Platelets are directly involved in hemostasis – a process that prevents blood loss. Thrombocytes release plenty of thread-like fibers that form blood clots. Platelets work hand in hand with fibrinogen, a protein that is easily released in the site of the cut and stops the bleeding. This prevents excessive blood loss that can lead to hypovolemic shock and even death.

Thermoregulation

Maintaining normal body temperature is one of the important functions of the cardiovascular system. The normal body temperature of a healthy adult is between 36.5-37.5°C. If the temperature is lesser than the normal, it is referred to as hypothermia and if it is higher than the normal, it is called hyperthermia. Deep organs like the brain, liver, and heart are the ones that generate most body heat.

During normal weather conditions, the heart typically releases only 4% of blood to the skin to keep it hydrated and within a normal temperature range. When the weather is too hot, the hypothalamus (body's thermostat) signal the heart to pump 48% of blood to the skin to keep it hydrated. The smooth muscles in the arterioles relax and cause vasodilation to allow more volume of blood flow to the skin. When the thermoreceptors detect a sudden drop in skin or blood temperature, the hypothalamus sends message to the smooth muscles to vasoconstrict. This helps reduce the flow of blood to the skin and maintain a normal core body temperature.

Fluid Balance

The cardiovascular system, together with nervous and endocrine systems, keeps the body's fluid levels within a normal range. Fluid balance is necessary to maintain an efficient flow of nutrients, gases, and electrolytes through the cells. Decreased fluid levels result in dehydration that can be life-threatening if not treated immediately. This is accompanied by an excessive loss of main electrolytes such as calcium, potassium, and sodium. Hyperhydration, on the other hand, is the result of excessive fluid intake that results in an increased amount of electrolytes in the cell.

This can cause swelling of the body tissues and different body organs. If the swelling happens in the brain, this will cause excessive pressure on the brain stem. This can result in seizure attacks, brain injury, coma, and death.

Changes in blood volume and tonicity can be easily detected by the kidneys and the osmoreceptors in the hypothalamus. Osmoreceptors are specialized receptor cells that detect sudden changes in blood dilution. These receptors can easily know if a person is dehydrated, hydrated, or overhydrated. The cardiovascular system responds by releasing hormones to act on different organ tissues to increase or decrease fluid production. Another way is by constricting or dilating blood vessels to increase or decrease the fluids that be lost by sweating.

Homeostasis

Homeostasis is a balanced state of all internal physical and chemical conditions performed by different body systems. This is a dynamic state of equilibrium that brings optimal functioning to an individual. This includes normal body temperature, fluid-electrolyte balance, normal blood pH and good blood sugar levels. All these variables are regulated by several homeostatic mechanisms, which altogether maintain and improve life.

All body systems contribute to homeostasis but the cardiovascular system provides the greatest contributions to overall body balance. The circulatory system works with the respiratory system to provide oxygen and nutrients to different body tissues and organs. The heart pumps the blood that goes to different regions of the body while the blood vessels serve as the passageways and communicating structures. The blood carries hemoglobin that contains the most important gas in the human body – oxygen. The circulatory system makes all bodily functions possible. Without it, all other body systems will not function and die. The heart, blood vessels, and the lungs work together to maintain homeostasis.

The Cardiac Cycle

The cardiac cycle is a series of events that happen during every heartbeat. This involves heart contraction (systole) and heart relaxation (diastole). Systole is the period of ventricular contraction that forces the blood to go into the aorta and pulmonary artery. Diastole, on the other hand, is the period of ventricular relaxation that causes the ventricles to be refilled with blood. The two atria and the two ventricles work harmoniously to ensure blood is efficiently throughout the body.

Cardiac output

The cardiac output is the measurement of the amount of blood pumped by one ventricle (stroke volume) in one full minute. This can be computed by multiplying the stroke volume by the heart rate (CO=SV x HR). The normal cardiac output is about 4.0-8.0 L/min.

Chapter 10. The Anatomy and Physiology of the Heart

The human heart is a cone-shaped, muscular organ of the cardiovascular system that is about the size of a closed fist. The base of the heart is positioned upwards and tapering down to the apex which is the bottom part. The normal weight of the heart is between 250-350 grams or 9-12 ounces while its dimensions are: 10-12cm length, 2.8-3.5 width, and 5-6.5 cm thickness.

Function

The two main functions of the heart are:

1.   To collect oxygenated blood from different body tissues and pump it to the lungs for oxygenation

2.   To collect oxygenated blood from the lungs and pump it to the systemic circulation

Aside from collecting and pumping blood to different tissues, the heart also helps bring waste products from the cells to the kidneys. These body wastes include nitrogenous wastes, salts, and excess fluid.

Another important function of the heart is pumping interstitial fluid from the blood to the extracellular space. The interstitial fluid contains nutrients, electrolytes, and dissolved gases that are needed to sustain life. The lymphatic system returns the excess interstitial fluid to the circulatory system via lymphatic vessels.

Location of the Heart

The human heart is located between the two lungs, at the level of T5-T8 (thoracic vertebrae) in the middle mediastinum, posterior to the sternum.   The   pericardium,   a double-membraned sac, surrounds the heart and attaches it to the mediastinum. The human heart rests on the top surface of the diaphragm and lies inside the protective thorax. The thoracic cavity or the thorax houses and heart and protects its delicate structures inside.

The front surface of the heart is situated behind the rib cartilages and sternum while the back surface lies near the vertebral column. The upper portion of the heart which serves as the attachment points for large blood vessels such as the aorta, pulmonary trunk, and venae cavae, is situated at the level of the 3rd costal cartilage. The apex, the lower tip of the heart is situated at the left side of the sternum, about 8-9 cm away from the midsternal line.

Main Parts of the Heart

The main components of the hearts are the heart chambers, valves, walls, vessels, and the conduction system. The right side of the heart collects unoxygenated blood from different body tissues and pumps it into the lungs. The left side, on the other hand, collects oxygenated blood from the lungs and distribute it to different body tissues. The left and right sides of the heart are divided by a wall of muscles called septums – atrial and ventricular. The atrial septum divides the left and right atriums while the ventricular septum divides the left and right ventricles.

The Heart Chambers

The heart has four chambers: right atrium, left atrium, right ventricle, and left ventricle. The two upper atria are called the receiving chambers while the two lower ventricles are referred to as the discharging chambers. The combination of the right atrium and right ventricle is sometimes called the right heart while the left atrium and left ventricle combination is referred to as the left heart.

The Right Atrium

The right atrium is the right upper collecting chamber of the heart. It receives unoxygenated blood from different body tissues through the inferior and superior vena cava. Contraction of the right atrium lets the blood pass to the right ventricle. The posterior part of the right atrium is made up of a smooth wall called the *sinus venarum,* while the anterior part is lined by parallel ridges of muscles termed as the pectinate muscle. The ostium, the inferior of vena cava, is located at the inferior border of the right atrium. The posterior wall of the right atrium receives the coronary sinus, the inferior vena cava, and the superior vena cava.

The Right Ventricle

The right ventricle receives unoxygenated blood from the right atrium and pumps it to the lungs for oxygenation. The pulmonary artery transports the unoxygenated blood to the lungs where it releases the carbon dioxide and picks up oxygen. Majority of the anterior surfaces of the heart is formed by the right ventricle. The *carneae,* the abundant coarse trabeculae, forms the walls of the right ventricle. The *conus arteriosus* also called the infundibulum is a smooth-walled artery that carries blood away from the ventricle. The component of the infundibulum that separates the right and left outflow tracts is the infundibular septum. The semicircular arch – the four muscle bundles separate outflow tract from other parts of the right atrium.

The Left Atrium

The left atrium receives the oxygen-rich blood from the lungs through the right and left pulmonary veins. The left atrium is located in the posterior part of the right atrium just above the left ventricle. The posteriolateral connections of the pulmonary veins (left and right) are the result of the venous structures absorbed into the left atrium. The atrial septum that is derived from the embryonic septum primum gives rise to the foramen ovale – a sealed valve flap.

The Left Ventricle

The left ventricle receives the oxygenated blood from the left atrium and distributes it to all the body tissues through the aortic artery. The left ventricle forms the majority of the heart's left lateral surface, and some parts of the inferior and posterior surface. Like the right ventricle, the carneae also characterize the walls of the left ventricle. However, the muscular ridges of the left ventricle are relatively fine unlike those of the right ventricle.

The Valves

The heart has four valves: tricuspid, mitral, pulmonary and mitral. These heart valves separate the heart chambers. The valves that separate the atria and ventricles are referred to as the atrioventricular (AV) valves. The two AV valves are the tricuspid valve and the mitral valve. The chordae tendinae, the cartilaginous connections from the papillary muscles, keep the heart from eversion prolapse. Meanwhile, the two semilunar valves, which are pulmonary and aortic, are located at the exit points of the right and left ventricles.

The Tricuspid Valve

The tricuspid valve is an atrioventricular valve that separates the right atrium and right ventricle. The unoxygenated blood is pumped from the right atrium via the atrioventricular orifice to the right ventricle. When the right ventricle contracts, the tricuspid valve prevents the blood from flowing back into the right atrium. The main parts of the tricuspid valve are the annulus, three sets of chordae tendineae, three papillary muscles, and three valvular leaflets.

The Mitral Valve

The mitral valves, also called bicuspid valve, is another atrioventricular valve that separates the left atrium and the left ventricle. It is named bicuspid valve because of its two cusps that are attached to the two papillary muscles. The oxygenated blood from the left atrium is pumped into the left ventricle through the left atrioventricular orifice. When the left ventricle contracts, the left AV valve or the bicuspid valve prevents the blood from flowing back to the left atrium. The mitral valve is composed of the annulus, two sets of chordae tendinae, two papillary muscles, and two leaflets.

The Pulmonary Valve

The pulmonary semilunar valve is situated at the base of the pulmonary artery. During ventricular systole, the unoxygenated blood is pumped from the right ventricle into the pulmonary arteries toward the lungs. During ventricular diastole, the pulmonary semilunar valve prevents the blood from going back to the right ventricle. The pulmonary valve consists of three symmetric, semilunar cusps.

The Aortic Valve

The aortic valve or the semilunar aortic valve is situated at the base of the aorta. It is not at all connected to the papillary muscles. The ventricular systole results to the movement of oxygenated blood from the left ventricle to the aortic artery. The oxygen-rich blood will go the systemic circulation and will be distributed to different body tissues. The aortic semilunar valve prevents blood from flowing back into the aortic artery during ventricular contraction. The aortic valve has also three symmetric, semilunar-shaped cusps like the pulmonary valve.

The Walls

The three layers of the heart wall are the endocardium (inner), myocardium (middle), and epicardium (outer). The pericardium, a double-membraned sac, surrounds these three heart layers.

The Endocardium

The endocardium is the innermost layer of the heart that is lined with simple squamous epithelium. This layer covers the 4 major heart chambers and the 4 major heart valves. The endocardium secretes endothelin, that regulates the contraction of the myocardium.

The Myocardium

The myocardium is the middle layer of the heart wall that mainly consists of cardiac muscles. These are striated muscle tissues that are surrounded by a framework of collagen. The cardiac muscles have a complex yet elegant swirling pattern that enables the heart to pump blood more efficiently.

The Epicardium

The epicardium is the outer covering of the heart. This is the thin, innermost layer of the pericardium where the coronary arteries lie. The epicardium surrounds heart structures and the roots of the great blood vessels.

The Pericardium

The Greek word pericardium means around (*peri*) the heart (*cardia*). The pericardium which serves as the outer covering of the heart is composed of two continuous layers that are divided by a potential space that contains the serous fluid – a lubricating fluid.

The pericardium is divided into two parts – the visceral pericardium and the epicardium. The visceral pericardium, also called epicardium, is the part of the pericardium that touches the heart. The surface of the visceral pericardium is covered by the mesothelium – a single layer of flat-shaped epithelial cells. Other parts of the epicardium include a single, thin layer of fibro-elastic connective tissue that supports the mesothelium, and a thick layer of adipose tissue that connects the fibroelastic layer to the myocardium.

The parietal pericardium, on the other hand, forms the outer border of the pericardium. This part of the pericardium does not touch the heart. The parietal pericardium contains serous (watery) and fibrous layers. These layers contain elastin fibers and collagen that give strength and elasticity to the parietal pericardium.

Great Vessels

Great vessels are the major blood vessels that carry blood toward and away from the heart. The five great vessels are the following:

The Superior Vena Cava

The superior vena cava is a 24-mm long vein that receives unoxygenated blood from the upper portion of the body. This main vessel is located in the anterior right portion of the mediastinum. The superior vena cava is the most common site of central venous access during the insertion of a central venous catheter or a peripheral central catheter.

The Inferior Vena Cava

The inferior vena cava is another large vein that receives deoxygenated/unoxygenated blood from the lower parts of the body. The walls of the inferior vena cava are so rigid to avoid flowing of blood to the lower portions of the heart. This major vein is formed by the junction of the left and right common iliac veins.

The Pulmonary Arteries

The pulmonary artery carries unoxygenated blood from the right ventricle to the lungs for oxygenation. The pulmonary trunk or the main pulmonary artery is the largest pulmonary artery of the heart while the arterioles are considered to be the smallest.

The Pulmonary Veins

The pulmonary veins are the veins that transport oxygen-filled blood from the lungs to the heart. The main pulmonary veins are the largest pulmonary veins that originate from the two lungs and drain to the left atrium.

The Aorta

The aorta is considered as the largest major artery of the human body. It originates from the left ventricle and ends to the abdomen where it branches into two smaller arteries – the common iliac arteries. The main function of the aorta is to transport the oxygenated blood into the systemic circulation.

The four sections of the aorta are:

• ***Ascending Aorta*** – It begins from the opening of the aortic valve and runs through the pericardial sheath and the pulmonary trunk. The twisting of these two blood vessels causes the aorta to start out posteriorly in the pulmonary trunk.

• ***The Aortic Arch*** - The aortic arch loops over the bifurcation of the pulmonary trunk and the left pulmonary artery. It is connected by the ligamentum arteriosum, a remain of fetal circulation that is destroyed a week after birth.

• ***The Thoracic Aorta*** – The thoracic aorta or the thoracic descending aorta gives rise to the subcostal and intercostal arteries and also to the inferior and superior left bronchial arteries.

- ***The Abdominal Aorta*** – The abdominal aorta starts from the aortic hiatus of the diaphragm to the bifurcation of the right and left iliac arteries.

The Conduction System

The sinus rhythm, or the normal rhythmical beating of the heart, is started and established by the sinoatrial (SA) node. The SA node is the heart's pacemaker which creates electrical signals that can travel through the heart. This causes contraction of the cardiac muscles. The SA node is located at the upper portion of the right atrium near the superior vena cava.

The electrical signals created by the SA node travel through the right atrium in an odd, radial way that can't be fully explained. The signals then travel to the left atrium via Bachmann's bundle and to the atrioventricular (AV) node.

The AV node is situated at the bottom of the right atrium in the AV septum – a muscle wall that divides the right atrium and the left ventricle. The atrioventricular (AV) septum is a part of the cardiac skeleton. This is a tissue inside the heart that does not allow electrical signals to pass through which forces the electrical impulses to pass through the AV node. The signal will travel to the bundle of His and the heart ventricles. In the ventricles of the heart, the Purkinje fibers – the specialized tissues, will then transmit the signals to the heart muscles.

Chapter 11. Blood And Blood Vessels

Blood is the red liquid component that flows inside the human body. It is made up of 92% water, and cells and plasma, a colorless fluid in which corpuscles or fat globules are suspended. On average, there is about 5 liters of blood circulating in the body.

Blood Functions

Blood is responsible for carrying out various crucial functions across the different bodily systems.

It transports nutrients digested from food intake and delivers oxygen from the lungs to the whole body.

It conveys hormones from the glands into the corresponding systems. It also picks up wastes from around the body and helps regulate body temperature

Composition Of Plasma

Plasma, one of the main components of blood, is made up of the following sub-components:

- 90% water
- 7-8% soluble proteins – three important protein types are albumins, globulins and clotting proteins
- 1% elements in transit
- 1% carbon dioxide

Proteins In Plasma

As mentioned in the previous section, there are three important protein types present in plasma:

- Albumin - Albumin comprises two thirds of the protein content of plasma. This protein type is produced in the liver with a main function of maintaining the osmotic balance between tissues and blood. Albumin also assists in the transport of vitamins across the body.

- Globulins - This protein type can be further classified as gamma, alpha and beta. Globulins aggregately assist the immune system in fighting off infections and illnesses.

- Clotting proteins - There are 12 substances that aid in the clotting process of blood. All these clotting proteins are also produced in the liver. Fibrinogen is the name of the main protein responsible for blood clotting.

Main Types Of Blood Cells

There are different types of cells that make up the blood:

Red blood cells

Red blood cells deliver oxygen from the lungs to the rest of the body and also pick-up waste released by the lungs. An example of these wastes is carbon dioxide. Red blood cells are able to carry out oxygen because they contain hemoglobin, a protein that binds to and releases oxygen.

Hemoglobin is the main component of red blood cells with each RBC containing 250 million hemoglobin. Oxyhcmoglobin refers to the interaction of hemoglobin, oxygen and iron and gives blood its red color.

Red blood cells look round and flat with an indented center. They are 7 – 8 micrometers in diameter. This appearance and size are what allows these cells to travel even through small capillaries.

Red blood cells are formed in the red bone marrow with the process of formation called erythropoiesis. There are 2-3 million RBCs produced by the human body per second making these cells 1000 times more plenty compared to white blood cells. The lifespan of an RBC is 120 days and they do not self-regenerate.

White blood cells

White blood cells detect and fight infections or foreign molecules in the body. Without them, the body would be vulnerable to every single bit of virus and bacteria it is exposed to.

White blood cells are larger than red blood cells. They are 10 – 14 micrometers in diameter and are translucent.

White blood cells can be further broken down into types, as follows:

- Basophils store and synthesize histamine important in protecting the body against allergies
- B & T-Cell Lymphocytes serve as markers for pathogens then subsequently destroys them
- Eosinophils kill parasites
- Monocytes are the biggest among the types of white blood cells. They push cells to work all together to defend the body against bacteria.
- Neutrophils are the most abundant type of white blood cells that mainly fight

infection. The neutrophils' process of fighting off bacteria is called phagocytosis. A neutrophils' lifespan ranges only from 12 to 48 hours.

Platelets or thrombocytes

These are small cells responsible for blood clotting. Blood clotting is necessary for not letting the body bleed to death whenever a person would have small cuts or bruises. They also carry nutrients from the food taken in such as sugars, fat, proteins, vitamins and minerals.

Platelets measure 1 to 2 micrometers in diameter. The human body produces 200 billion platelets per day. This process of platelet production is regulated by the hormone Thrombopoietein. A platelet's lifespan ranges from 8 to 10 days.

All these type of blood cells are manufactured in the bone marrow.

Blood Vessels

Blood vessels are like a web of small tubes spread throughout the body responsible for carrying blood. Collectively, blood vessels compose the vascular system. Peripheral vascular system is the term to refer to all arteries and veins that are outside the heart and head.

Note however, that blood vessels do not have peristalsis. It is the pressure from the heartbeat that propels blood through them.

The Different Types Of Blood Vessels

Arteries

Arteries are the largest and strongest type of blood vessel that carries blood pumped from the heart. They can expand and contract to lower or increase blood pressure according to the body's needs. Arteries carry oxygenated blood from the lungs to all the other organs.

For blood in all arteries except in the pulmonary artery, the hemoglobin component is 95-100% saturated with oxygen.

Veins

Veins carry deoxygenated blood from the organs back to the heart. For blood in all veins except in the pulmonary vein, the hemoglobin component is 75% saturated with oxygen. There are different kinds of veins:

- Subclavian vein – a paired large vein responsible for draining blood from the upper extremities back into the heart
- Jugular veins – veins that transport deoxygenated blood from the brain back into the heart

- Renal veins – carry blood filtered by the kidney
- Iliac vein – drains blood from the pelvis and lower limbs
- Venae Cavae – two largest veins in the body
- Venules – very small veins that connect to capillaries

Capillaries

These are the tiniest type of blood vessels that connect arteries and veins. Capillaries allow blood to come close to the different body tissues in order for the exchange of water and chemicals to take place. They are made up of a single layer of endothelial cells with a supporting subendothelium (basement membrane and connective tissue)

Blood vessels cover almost the entire body except for structures such as the cartilage, epithelium and the lens and cornea of the eye. These structures are called avascular structures, meaning not having blood vessels.

Other structures and types of blood vessels include

- Anastomosis – is the network of blood vessels connected to form a region of diffuse vascular supply. This provides an alternative blood flow in case of blockages.
- Valves – this is part of blood vessels that prevent backflow of blood against gravity
- Sinusoids – are the extremely small blood vessels located within the bone marrow, spleen and liver
- Arterioles – these are small branches of arteries that connect to capillaries

Size Of Blood Vessels

The average diameter of blood vessels vary widely depending on the type. An aorta is 25 millimeters thick on average while a capillary is normally just 8 micrometers.

The size of blood vessels may also vary depending on external factors. Vasoconstriction refers to the normal narrowing of blood vessels controlled by a combination of the vascular smooth muscles, paracrine factors, hormones and neurotransmitters. Vasodilaton, on the other hand, refers to the normal widening of the blood vessels. It typically occurs in response to triggers like low oxygen levels and temperature increase.

Layers Of Arteries And Veins

Each artery and vein comprises several layers.

Tunica Intima – this is the inner and thinnest layer. Tunica intima is composed of a single layer of flat cells called simple squamous epithelium glued together by a polysaccharide intercellular matrix and surrounded by a thin layer of subendothelial connective tissue interlaced with circularly arranged elastic bands called internal elastic lamina.

Tunica Media – this is the middle layer. Tunica media consists of circularly arranged elastic fibers, connective tissues and polysaccharide substances. This layer is thicker in arteries than in veins and is the thickest layer in arteries.

Tunica Media in arteries are rich in vascular smooth muscle to control the caliber of the vessel. This layer is separated from the outermost layer by a thick elastic band called external elastic lamina.

Tunica Adventitia – this is the outermost layer. The tunica adventitia is the thickest layer in veins. It is entirely made up of connective tissues and contains nerves.

Blood Flow

The unit of measurement for blood pressure is millimeters of mercury (mmhg). Systolic is the term for the high pressure wave of blood due to contraction of the heart while diastolic is the low pressure wave of blood.

The average systolic pressure in the arterial system is 120mmhg while the average diastolic pressure in the arterial system is 80mmhg.

Blood pressure in the venous system is constant and rarely exceeds 10mmhg.

Vascular resistance is the condition wherein the vessels away from the heart oppose the flow of blood. This is caused by combined factors of blood viscosity, vessel length and radius. Blood viscosity refers to the thickness of blood and its resistance to flow.

Blood Types

A person's blood type is determined or classified by the presence or absence of crucial substances in them. The type of blood that a human will have is inherited genetically.

The crucial substances used to classify blood are the two antigens A & B and the protein Rh. Antigens are substances that trigger an immune system response opposing a foreign body such as bacteria. The presence or absence of Rh in a person's blood is denoted by the symbol + or -.

The 8 blood types of humans are:

A+

A-

B+

B-

O+

O- : Called the universal red cell donor. This is the most common blood type.

AB+

AB-: Called the universal plasma donor

The A group are those blood types that have A antigen on red cells and B antigen in the plasma.

The B group are those blood types that have B antigen on red cells and A antigen in the plasma.

The AB group are those blood types that have both the A and B antigen on the red cells and none on the plasma.

The O group are those blood types that have neither A nor B antigen in the red cells but have both antigens present in the plasma.

Genetics Of Blood Types

As mentioned in the section Blood Types, the type of blood that a person will have is dictated by genetics. This section will list all possible parent blood type combinations and the possible resulting blood types in the offspring.

Parent A + Parent B = O, A, B or AB offspring

Parent AB + Parent AB = A, B or AB offspring

Parent AB + Parent B = A, B or AB offspring

Parent AB + Parent A = A, B or AB offspring

Parent AB + Parent O = A or B offspring

Parent B + Parent B = O or B offspring

Parent A + Parent A = O or A offspring

Parent O + Parent B = O or B offspring

Parent O + Parent A = O or A offspring

Parent O + Parent O = O offspring

Chapter 12. The Lymphatic System And Immune System

The lymphatic system is a network of organs and tissues that help in getting rid of toxins, wastes and other unwanted particles in the body. The lymphatic system and the action of lymphocytes are called as the adaptive immune response. These are highly specific and long-lasting responses to pathogens. It consists of lymph, lymphatic vessels, red bone marrow, and a number of body structures and organs containing lymphatic tissues. Lymph is the fluid that flows through the lymphatic system which contains infection-fighting white blood cells in the body.

Lymphatic Vessels

Networks of tiny, thin-walled micro vessels known as lymphatic capillaries are one of the structures which belong to the lymphatic system.

<u>Lymphatic capillaries</u>

Lymphatic capillaries are located in almost every tissue in the body and in the spaces between cells. They are also interlaced among the venules and arterioles of the circulatory system. Its structure is similar to that of a blood capillary, but functions differently. The lymphatic capillary is where interstitial fluid enter going to the lymphatic system to become lymph fluid.

Interstitial fluids are the fluid found in between the cells in all the body tissues. It is responsible for the transport of nutrients and wastes products between cells and blood capillaries.

Its composition and properties differs between organs and tissue development.

On the other hand, lymph fluid is the derivative of the interstitial fluid. It returns proteins and excess interstitial fluid to the bloodstream, and transports fats from the digestive system to the blood. Lymph's composition continually changes because of its continual exchange of substance with the interstitial fluid. The usual color of lymph is clear to pale yellow, but its color in the human digestive system is milky white because of its lipid content. This milky white lymph is called chyle and is rich in triglycerides.

The difference between interstitial fluid and lymph fluid is their location. The interstitial fluid is found between cells, while lymph is located within the lymphatic tissue and lymphatic vessels. Interstitial fluid is formed in the filtered blood plasma from the blood capillary walls, while lymph is formed after interstitial fluid passed into the lymphatic vessels.

Lymphatic capillaries join together to form larger tubes call lymphatic vessels. They are described as small veins that have thin walls and more valves which are complementary to the cardiovascular system. They are committed to the propulsion of the lymph fluid from the lymph capillaries.

Lymph vessels that send lymph to a lymph node are called afferent lymph vessels, and those that receive lymph from a lymph node are called efferent lymph vessels. Lymph that was received from a lymph node are already filtered and may be transported to another lymph node, may go to a larger lymph duct, or may be returned to a vein. Lymph node is a bean-shaped organ which filters the lymph before it is returned to the circulatory system. They serve as checkpoints for the immune system and serves as patrol for protection from infections.

Lymphatic Trunks

Efferent lymph vessels that merge together will form what is called as lymphatic trunks, which in turn drains into one of the lymph ducts. The main trunks in the body include the bronchomediastinal trunk, intestinal trunk, lumbar trunk, subclavian trunk, and the jugular trunk.

1. Bronchomediastinal trunk - Drains lymph from the lungs, heart, and the thoracic wall.
2. Intestinal trunk - Drains lymph from interstines, stomach, spleen, pancreas, and some parts of the liver.
3. Lumbar trunk - Responsible for draining lymph from kidneys, adrenal glands, lower limbs, abdominal wall, and from the pelvic viscera wall.

4. Subclavian trunk - Drains lymph from the upper limbs
5. Jugular trunk - Drains lymph from the head and neck.

The lymph passes from lymph trunks going to the thoracic duct and to the right lymphatic duct, then depletes into the venous blood. The right lymphatic duct receives lymph from the upper right side of the body only, while the thoracic duct receives lymph from the other parts of the body.

Lymphoid Tissue

The lymphoid tissue is an important part of the body's defense system. This is where lymphocytes are usually found. Lymphocytes are one type of white blood cells that are also considered as one of the body's main types of immune cells. They constantly patrol the body in pursue of looking for pathogens, which they can eradicate and eliminate. Lymphocytes can be found on human organs and cells in the body.

Lymphoid tissues are classified as primary and secondary in nature. The primary lymphoid tissues are the thymus and the bone marrow. It is where lymphocyte matures, initially differentiates, and where they develop their central tolerance. It is also where most lymphocytes-recognizing self antigens are removed, and only cells recognizing foreign antigens are permitted to mature. Secondary lymphoid tissues include the tonsils, lymph nodes and the spleen, which provide sites for both the gathering of the lymphocytes and the accumulation of antigen.

Primary Lymphoid Tissues

Bone Marrow

The yellow bone marrow is mostly made out of fat cells and is a site of energy storage, while the red bone marrow is a loose collection of cells where hematopoiesis occurs. The maturity of B cells happens in the red bone marrow.

Thymus

The thymus gland is found in the mediastinum of the aorta of the heart and sternum. This is divided by connective tissue capsules, which are called trabeculae to create lobules. Its outer region is called cortex, which is densely packed epithelial cells, macrophages, and dendritic cells. Its middle part is called medulla, which contains less dense thymocytes, dendritic cells, and epithelial cells. Dendritic cells assist in the maturation process of lymphocytes, while the epithelial cells help educate the pre-T cells. The thymus, as a whole, stores immature lymphocytes and prepares them to be active T cells to help destroy cancerous cells.

Secondary Lymphoid Tissues

Tonsils

Tonsils are large clusters of lymphatic cells located in the pharynx. It is where samples of bacteria and viruses that enter through the mouth and nose are collected, so that the body can create proper antigens

Spleen

The spleen is also known as the filter of the blood because it removes microbes and dead red blood cells in the body. It is located on the left side of the body just above the kidney. It is a fragile organ since it does not have a strong capsule. Its color is dark red because of its extensive vascularization. It functions as the location of immune response to blood-borne pathogens. When the spleen detects dangerous organisms in the blood, it creates lymphocytes or white blood cells, which defends the body against these harmful elements. These dangerous organisms are fought by the white blood cells by producing antibodies, which will kill them, and thus stop infections from spreading in the body.

Lymph Nodes

here are 600 lymph nodes in the body, which swells in response to infection. Lymph nodes are small bean shaped structure found in the different parts of the body. They are often referred to as glands, but this is not true because they do not form the part of the endocrine system. It is in the lymph nodes where immune cells assess foreign organisms such as bacteria, fungus, and viruses to see if they are harmful for the body. This happens after they receive the lymph fluid from the afferent lymph vessels, and filters it with the help of the lymphocytes. It is in the lymph nodes that the lymphocytes first encounter the pathogens, communicate with each other, and set off their defensive response.

Functions of the lymphatic system

1. It balances the amount of fluid between the blood and body tissues, also known as homeostasis.

2. It is a part of the immune system of the body since to helps the body in preventing bacterial and infectious growth by fighting foreign organisms.

3. It facilitates the absorption of fats and fat-soluble nutrients in the digestive system.

The Immune System

The immune system includes all the cells and tissues that carry out immune responses. Immunity is the ability of the body to fight off diseases through our natural body defenses. More information on the immune system can be found in Chapter 20.

The body's first line of defense includes: physical barriers, such as the skin; toxic barriers, such as the acidic contents of the stomach; and the presence of friendly bacteria in the body. The leukocyte or white blood cells are the main actors in the immune system. The two main types of leukocyte are the Phagocytes and the Lymphocytes.

Phagocytes

These cells protects the body by eating harmful foreign organisms, viruses and bacteria in the body. These cells surround and absorb these pathogens and break them down. There are many different types of phagocytes including:

1. Neutrophil – These are the most abundant type of phagocyte, which can immediately attack all antigens since they are not limited to specific areas of the body.

2. Monocytes – Monocytes are the largest type of phagocytes, which are influenced by the process of adaptive immunity.

3. Macrophages – These are formed in response to an infection, and can recognize and destroy their target cells. They play a critical role in the innate and adaptive immunity by recruiting immune cells called lymphocytes.

4. Mast Cells – These are tissue cells of the immune system which releases histamine and other substances during inflammatory and allergic reactions.

Lymphocytes

Lymphocytes originate in the bone marrow and helps the body to fight pathogens. But before anything else, it's important to note that there are two types of immunity: first is innate immunity and the second is adaptive immunity.

Innate immunity is a fast immune defense response, which provides the first line of defense of the body against infections. It is also known as the nonspecific body defense. It is comprised by different cells such as natural killer cells, monocytes, eosinophils, macrophages, tor-like receptors, and a series of soluble mediators.

Adaptive immunity, on the other hand, is also referred to as specific immunity. This refers to the system that mainly attacks specific invaders, and is activated after an exposure to an antigen. Antigens are antibody generators, which are produced when foreign substances enters the body.

There are two types of adaptive immunity namely: cellular immunity and humoral immunity. Cellular immunity is where the T lymphocyte eliminates infected cells and provides help to other immune responses. Humoral immunity is when B lymphocyte secretes antibody molecules to neutralize the pathogen outside the cells to prevent diseases.

Lymphocytes are the primary cells of adaptive immune responses. Its two basic types are the T cells and the B cells. Both are initially developed in the bone marrow, but mature on different areas.

B cells and T cells are named based on where they mature. B stands for bursa equivalent – the red bone marrow, and T stands for the thymus gland. Before T cells and B cells leave their maturation area, they develop the ability to carry out adaptive immune responses called immunocompetence. Immunocompetence is where B cells and T cells begin to make several distinctive proteins that functions as antigen receptors, which are inserted into their plasma membranes. Antigen receptors are basically antibody proteins that are not secreted, but are anchored to the B cell and T cell membranes.

These cells are capable of recognizing foreign antigens, and are able to create immunological memory. This allows them to recognize the pathogens, which they have encountered before.

They are activated in secondary lymphatic organs and tissues such as the spleen, lymphatic nodes, and lymphatic nodules. When activated, they form clones of cells that can recognize specific antigens. They are also identical with each other, but can be distinguished by the molecules they secrete, and by their surface protein markers.

B cells primarily function by producing antibodies, which is why they are also called immune cells. An antibody is a group of proteins that binds to antigens – pathogenic molecules. Antigens are chemical structures on the surface of a pathogen, which binds to B lymphocyte or T lymphocyte antigen receptors. Activated B cells are also known as plasma cells. These plasma cells secrete antibodies that inactivate pathogens, and mark them for destruction.

Unlike B cells, T cells do not secrete antibodies. T cells can destroy infected cells with intracellular pathogens. There are two types of mature T cells namely, the helper T cells and the cytotoxic T cells. Helper T cells are also known as CD4 T cells since their plasma membranes include CD4 protein. Cytotoxic T cells are also known as CD8 T cells since they contain CD8 protein. Most of the time T cells are inactive, and the first signal in activation of a T cell would be the T cell receptors with CD4 and CD8 proteins. T cells only become activated if it binds to foreign antigen, and at the same time receives the process of costimulation. Costimulation is the second signal for T cell activation. It may prevent immune responses from occurring accidentally. Also, recognition without costimulation can lead to prolonged anergy - a state of inactivity in both the T cells and B cells.

The immune system is very vital to everyone's survival. Each one has different immune systems, which becomes stronger as time pass by because of numerous exposure to pathogens. Several diverse systems and cells work together all throughout the body to fight off pathogens and clear up dead cells to ensure that the body is safe and healthy.

Chapter 13. The Respiratory System

A human respiratory system is a group of organs that function together to store oxygen and eliminate carbon dioxide. The human body needs a constant supply of oxygen to provide nourishment to different body cells and tissues. It also has to expel carbon dioxide, a toxic gas that can be fatal if accumulated inside. The three main parts of the human respiratory system are the airways, the lungs, and the muscles of respiration.

Functions of the Respiratory System

The main function of the respiratory system is to oxygenate the blood and eliminate carbon dioxide. This will be accomplished through respiration. Respiration is the movement of oxygen to the lungs and the release of carbon dioxide from the lung tissues. The four processes of respiration are:

1.   Pulmonary Ventilation

This is more commonly known as breathing. Pulmonary ventilation is the continuous movement of air to and from the lungs. This process keeps the flow of gases in the lungs clean and safe to use.

2.   External Respiration

This happens when oxygen from the lungs moves to the bloodstream and carbon dioxide from the blood goes to the lungs. This gas exchange occurs by diffusion, which refers to the process in which a substance moves from an area of higher concentration to one with lower concentration.

3.   Transport of Respiratory Gases

This is the process of transporting oxygen from the lungs to tissue cells and of carbon dioxide from tissues to the lungs. The cardiovascular system is responsible for this action. The blood serves as the transporting fluid for this gas transport.

4.   Internal Respiration

This is the movement of oxygen (oxygen diffusion) from the blood into cells and interstitial fluid. Carbon dioxide and wastes, on the other hand, are diffused from the cells and interstitial fluid to the blood.

Other functions of the respiratory system include:

- Filtration and humidification of the air that you breathe
- Maintains homeostasis, or balance inside the body
- Aids in *phonation*, or creation of sound

- Plays an active role in *olfaction* or smelling

The Parts of Respiratory System

The respiratory system is divided into these two major areas:

The Conducting Zone/The Airways

Consists of organs and structures that provide a continuous passageway for air to travel in and out of the lungs. These include the nasal cavity, larynx, pharynx, trachea, bronchi, and bronchioles.

The Respiratory Zone

Consists of lung structures – terminal bronchioles, alveoli, and alveoli ducts that are responsible for gas exchange. These lung structures facilitate Oxygen diffusion in the lung capillaries in exchange for Carbon Dioxide gas.

The Conducting Zone/Airways

The conducting zone is also called the upper respiratory tract. The airway structures include the nose, pharynx, larynx, trachea, bronchi, and bronchioles. These organs collect, filtrate, humidify, and purify the incoming atmospheric air before it reaches the lungs. The following are the different airway organs and structures of the respiratory system:

Nose and Nasal Cavity

The nose is the outside protuberance of the nasal cavity – the interior of the nose. The nose is an external organ of the upper respiratory tract. It is the protruding part of the face that holds the nostrils. Its shape is formed by the nasal septum and the ethmoid bone. The *nasal septum* is the cartilaginous part of the nose that separates the nostrils. Meanwhile, the *ethmoid bone* separates the brain from the nose. This bone is mostly responsible for the shape and structure of the nasal cavities.

The nose warms and humidifies inhaled air that enters the lungs. It keeps the inhaled air humid and in good condition. The nose also has a group of specialized cells for smelling. This is more of a nervous function than a respiratory function.

The nasal cavity is divided into left and right canal by the midline nasal septum. This is a thin, bony, and cartilaginous wall that provides support to the nose. The nostrils or nares filter the air that goes inside the nasal cavity. The sticky mucus lubricates the air and traps foreign bodies that can harm other respiratory organs. The mucus contains lysosome enzymes that destroy pathogens chemically.

The nasal cavity is composed of nasal hairs, cilia (microscopic hairs), and mucous membranes. They trap the dust and pathogens from entering the pharynx and lung areas. Sneezing is the irritation of the nasal mucous that helps expel foreign bodies.The olfactory receptors are located in the superior part of the nasal cavity, just below the ethmoid bone. The conchae are mucosa-covered projections that increase the air turbulence in the cavity. The palates located at the base of the nasal cavity separates the oral cavity from the nose.

Paranasal Sinuses

These are air-filled spaces that come in pairs that surround the nasal cavity. Paranasal sinuses maintain the normal weight of the human skull and give resonance to the human voice. These also protect the face from injury and help humidify inhaled air. There are four different paranasal sinuses:

1. Maxillary sinuses – surround the nasal cavity
2. Frontal sinuses – above the eyes

3. Ethmoid sinuses – between the eyes
4. Sphenoid sinuses – at the back of the ethmoid bone

Pharynx

The pharynx, more commonly called as throat, is a 13-cm long muscular passageway that allows food and air to enter the system. The pharynx is divided into three regions - the nasopharynx, oropharynx, and laryngopharynx.

- The Nasopharynx

This is the upper region of the pharynx. It connects the nasal cavity to the throat. The nasopharynx is covered with ciliated pseudo-stratified squamous cell epithelial tissues. The nasopharynx contains adenoids (pharyngeal tonsils) that aid in producing T-lymphocytes for the immune system after birth.

- The Oropharynx (Mesopharynx)

This is the middle region of the pharynx. It is located in the middle of the oral cavity, below the nasopharynx, and above the laryngopharynx. The epiglottis is located between the oropharynx and the laryngopharynx. This is a flap of cartilage that immediately closes when a person is swallowing foods. This is to ensure that the food goes straight to the esophagus and does not enter the trachea. The mesopharynx also contains the palatine tonsils. These are not ciliated, masses of lymphoid tissues found on the walls of the oropharynx. These are removed in people with enlargement or infection.

- The Laryngopharynx (Hypopharynx)

The laryngopharynx is the lower region of the pharynx. It connects down to the esophagus and trachea. The hypopharynx divides the respiratory and digestive pathways. The hyoid bone separates the oropharynx from the laryngopharynx. The laryngopharynx does not contain any type of tonsils, unlike the two pharynx regions.

Larynx

The larynx or voice box is a short airway structure that connects laryngopharynx to the trachea. It is located just below the hyoid bone and above the trachea. Its interior consists of three regions – the *supraglottis*, *glottis*, and the *subglottis*. The glottis contains the two pairs of mucosal folds: the false vocal folds (vestibular) and the true vocal folds. The true vocal folds are responsible for sound production while the vestibular (false) folds are for resonance.

The vocal folds, or true vocal folds vibrate with expelled air to produce speech and vocal sounds. The vibration speed and tension of the true vocal folds can be adjusted to produce different voice pitches. The slit-like passageway structure that lies in the middle part of the vocal folds is called the glottis.

During swallowing, the larynx is covered by a cartilaginous part called epiglottis. Just below the epiglottis lies the thyroid cartilage, the largest hyaline cartilage that is enlarged in male adults. This is more commonly known as Adam's apple.

The primary function of the larynx is phonation or voice production. The vocal folds open when a person is breathing and close when he is swallowing or during phonation. Aside from phonation, the larynx also helps prevent aspiration or choking. When a person swallows, the larynx automatically closes and moves upward. This will cause the epiglottis to close the trachea. During coughing, the larynx closes to prevent foreign gases from entering the lungs. Lastly, the larynx can increase the Oxygen supply when a person suffers from the difficulty of breathing. It opens its folds wider to allow more airflow into the lungs.

Trachea

The 10-12cm long trachea or the windpipe extends from the base of the larynx to the level of 5th thoracic vertebra. Its main function is to clear the passageways of air that goes into and out of the lungs. The walls of the trachea are made of C-shaped hyaline cartilage rings and are lined with pseudostratified ciliated columnar epithelium. These rings remain open to allow air to enter the trachea. The ciliated epithelium produces mucus which traps foreign substances such as dust and other particles that can harm the lungs.

It also takes an active part in humidifying the air before it goes to the lungs. The trachea is a part of the anatomical dead space. This is the space in the airway that is not included in the alveolar gas exchange process. The anatomical dead space is a part of upper respiratory airways.

The Respiratory Zone

The respiratory zone is also referred to as the lower respiratory tract. These are the parts of the respiratory zone:

Bronchi and Bronchioles

At the base of the trachea, the airway splits into two primary bronchi – the left and right bronchi. These extend to the apex of the lungs and branch into smaller secondary bronchi.

The secondary bronchi transport air into different lobes of the left and right lungs. In the lobes of the lungs, the secondary bronchi split into much smaller branches known as the tertiary bronchi. These split further into tiny bronchioles that spread throughout the lungs. A single bronchiole splits into much smaller branches called terminal bronchioles. These are about 0.5-0.9 mm in diameter and their main function is to transport unoxygenated air to the alveoli.

The bronchi and bronchioles carry air from the trachea to the lungs for oxygenation. These airway structures are packed with elastin and smooth muscles that allow them to be more contractile and flexible. The smooth muscle tissues in the bronchioles help regulate airflow in the lungs. During strenuous activities, the smooth muscles relax to facilitate bronchial dilation. When the airway is dilated, the airflow resistance will be decreased to let air pass into and out of the lungs. When resting, the smooth muscles contract to avoid hyperventilation.

The Lungs

The cone-shaped lungs are the primary organs of the human respiratory system. The pair of lungs is located inside the thoracic cavity lateral to the heart and inferior to the diaphragm. The main function of the lungs is to transfer oxygen from the atmosphere into the bloodstream and to release carbon dioxide from the blood into the atmosphere. The apexes of the lungs extend to the root of the human neck just above the level of the sternum. The medial part of the lungs points towards the mid-chest, and lie against the great vessels, heart, and carina.

The hilum or the central recession of both lungs allow the airways and blood vessels to pass into and out of the lungs. The pulmonary pleurae – outer parietal and inner visceral, are two serous membranes that surround the lungs. The outer parietal pleura lines the wall of the rib cage while the inner visceral pleura lines the surface of the lungs. The pleural cavity, the space between the parietal and visceral pleurae, contains the lubricating pleural fluid. The left and right lungs are divided into different lobes by the infoldings of the pleura as fissures. These are two folds of pleura that section the lungs and aid in lung expansion.

The Right Lung

The right lung is divided into three lobes – upper, middle, and lower, by two fissures – oblique and horizontal. The horizontal fissure separates the upper lobe from the middle lobe. The point of division starts from the lower oblique just below the posterior border of the lung, to the anterior border in line with the sternal end of the 4th costal cartilage. The oblique fissure, on the other hand, divides the lower and middle lobes of the right lung. The lower oblique fissure is almost in line with the oblique fissure of the left lung.

The right lung's mediastinal surface is surrounded by several nearby structures. The azygos vein is situated in the anterior part of the hilum and just above this vein lies the superior vena cava and the right brachiocephalic vein. Behind the azygos vein near the top of the right lung is a groove for the brachiocephalic artery. The esophageal groove lies in the posterior part of the pulmonary ligament and the hilum. Below the esophageal groove lies the inferior vena cava groove.

The Left Lung
The left lung is divided into upper and lower lobes by the oblique fissure. The left lung does not have a middle lobe but it has a projection in the upper lobe called the lingula or the "little tongue". This projection serves as an anatomic parallel to the middle lobe of the right lung.

The lingula has two bronchopulmonary segments – superior and inferior.

The large cardiac impression where the heart lies is situated in the mediastinal surface of the left lung. Just above the hilum lies a curved groove for the aortic arch and below the hilum is a groove for the descending aorta. The left subclavian artery is located in a groove from the arch to the apex. The shallow groove for the left brachiocephalic vein sits near the edge of the left lung while the esophageal groove lies at the base of the lung.

The lungs of the human body are the most important organs of the respiratory system. The exchange of oxygen and carbon dioxide takes place in the lungs. The lungs also transport oxygen from the air to the blood and release carbon dioxide from the bloodstream to the air. Aside from gas exchange, other lung functions include:

- Acid-Base (pH) Balance - An increased amount of carbon dioxide can make a person acidic. When the lungs detect too much carbon dioxide in the body, they increase the ventilation rate to expel unwanted CO_2 gases.

- Filtration - The lungs can filter air embolism and small blood clots in the human brain and body.

- Infection Control - The lungs can excrete Immunoglobulin A, a type of antibodies that protects the lungs against infection.

- Blood Reservoir - The lungs can store up to 1, 000 ml (1L) of blood at its normal condition.

The Alveoli

The respiratory bronchiole gives rise to the alveolar sacs that contain the alveoli. These are tiny, hollow cavities inside the lung parenchyma that facilitates oxygen-carbon dioxide gas exchange. The alveoli have extremely thin walls to facilitate a faster rate of diffusion. The alveoli also contain an alveolar macrophage and the pneumocytes. Alveolar macrophages are phagocytic tissue cells that play an essential immunological role. They remove wastes and other substances that are deposited in the alveoli such as loose red blood cells.

The pneumocytes are also known as the Type I and the Type II of alveolar cells. Type I cells are also known as the flat squamous epithelial cells that form the alveolar wall structure. They have thin walls that are designed to facilitate fast and efficient gas exchange. The squamous cells also form the alveolar septa that separate the alveoli. Type II cells have a cuboidal shape and are found in the corners of the alveoli. They are typically larger than the Type I cells. They also produce lung surfactants and secrete epithelial lining fluid.

The Muscles of Respiration

The lungs are surrounded by muscles of respiration that aid the respiratory organs to breathe in and out air. The muscles of respiration are the diaphragm and the intercostal muscles.

Diaphragm

The diaphragm, the principal muscle of respiration, is a thin sheet of skeletal muscle that forms the base of the thorax. This is a thin, dome-shaped muscle that divides the abdominal and the thoracic cavity. The inhalation process causes the diaphragm to contract and compress the abdominal cavity. Also during inhalation, the diaphragm assists the ribs to move upward and outward. This is to facilitate the expansion of the thoracic cavity and push air into the lungs. Relaxation of the diaphragm causes elastic recoil of the thoracic wall to expel air from the lungs.

Intercostal Muscles

The intercostal muscles are a group of respiratory muscles that are attached to the middle part of the ribs. These muscles assist the diaphragm in compressing and expanding the lungs. Intercostal muscles are divided into two groups – the internal and external intercostal muscles. The internal intercostal muscles compress the thoracic cavity to force air to move out of the lungs. The external intercostal muscles elevate the ribs to allow volume expansion in the thoracic cavity to assist inhalation.

The Mechanics of Breathing

Breathing is a physiological process that allows atmospheric air to move into and out of the lungs. This is also called respiration. The body cells require a constant supply of oxygen to function efficiently. Deficiency in oxygen can lead to severe pathological consequences and different illnesses. Passive diffusion also called bulk flow, is a process of gas movement from an area of higher concentration to an area of lower concentration. Diffusion is a natural process that requires no amount of energy.

Inhalation

The inhalation process is started by the diaphragm and assisted by the external intercostal muscles. The normal respiration rate of a human is 12-20 breaths per minute. During normal inhalation, the diaphragm contracts, the rib cage expands, and the abdominal contents move downward. All these results in increased thoracic volume and negative pressure in the thoracic cavity. The inhaled atmospheric air moves from high-pressure zones to low-pressure zones. The contraction of the diaphragm allows air to travel through the airways. These airway structures, also known as the conducting zone, filter, warm, and humidify the air before it goes to the lungs. During hyperventilation (more than 30 breaths/min) and respiratory failure, the accessory muscles of respiration help the diaphragm and intercostals sustain or increase the respiratory rate. Accessory muscles of respiration include the sternocleidomastoid, scalene muscles of the neck, latissimus dorsi, and platysma.

Exhalation

Exhalation is considered a passive process. During exhalation, the chest is depressed and the rib cage is descended to let the air move out of the lungs. Also during exhalation, the dome-shaped diaphragm is elevated, the intrapulmonary pressure is increased, and the lungs recoil to a minimal volume. The elastic recoil of the lung is caused by the surface tension on the epithelium of the lungs. During forced or active exhalation, the expiratory, abdominal, and intercostal muscles work together to create pressure that forces air to move out of the lungs. Forced exhalation is performed to measure airway health and check airway obstruction.

The diffusion process enables oxygen to move from the alveoli to the blood capillaries. In the bloodstream, the hemoglobin collects the oxygenated blood. The oxygen-rich blood travel to the heart. The heart then pumps the blood through the arteries. These arteries bring the oxygenated blood throughout the body.

The oxygen from the blood is released from the hemoglobin and goes back to the cells. The carbon dioxide is absorbed by the capillaries and dissolved by the plasma. The carbon-dioxide-rich blood returns to the heart through blood-collecting veins. The blood is then pumped to the lungs for oxygenation. The alveoli eliminate carbon dioxide through the process of exhalation.

Chapter 14. Digestive System

The body needs energy to survive. It needs energy so it can function properly and stay healthy. It needs energy to grow and repair itself. The body gets this energy from food.

The digestive system turns food into nutrients through the process of digestion. It breaks down food into small molecules that the body can absorb and use to survive.

Digestion also produces waste matter. It is part of the digestive system's function to dispose of and eliminate residue or waste.

The digestive system consists of the digestive tract – also referred to as the gastrointestinal tract or the GI tract, and the gallbladder, pancreas, and liver.

The digestive tract is made up of several organs or parts connected in a long and twisting tube that work together to help the uniquely designed digestive system to perform its work.

The Organs of the Digestive System

The digestive system consists of the following organs:

Mouth

Digestion starts with the mouth. The process begins as soon as the first bite of food enters the mouth.

When a person chews or masticates food, the teeth cut, grind, and break the food down into smaller pieces which the body can digest more easily. The salivary glands in the mouth secrete saliva which contains enzymes and other substances that help to further break down the food. The tongue helps mix the saliva and food and, with the palate (or the roof of the mouth), moves the food to the throat (pharynx) and the esophagus as a person swallows.

Throat

The food passes from the mouth to the throat and to the esophagus.

Esophagus

The esophagus is found near the windpipe or the trachea in the throat. It is a muscular tube that extends to the stomach. In a process called peristalsis, the esophagus goes through a series of contractions to deliver the food to the stomach.

The lower esophageal sphincter (LES) is a ring-shaped valve found in the esophagus just before the part where the digestive tract opens to the stomach. The LES unlocks to move the food into the stomach. It then closes to prevent the food from coming back up.

If the lower esophageal sphincter does not function properly, a person is likely to experience GERD (Gastroesophageal reflux disease), a digestive disorder that causes regurgitation, the feeling of acid or food coming back up into the esophagus or mouth.

Stomach

The stomach is the muscular organ that keeps the food and further breaks it down into a more usable form. With the help of the powerful enzymes and acids that it excretes, the stomach grinds and mixes the food to the consistency of a paste. It then moves the food to the small intestine.

Small intestine

The small intestine, referred to as the digestive system's "work horse," is a long muscular tube made up of three parts, the duodenum, ileum, and jejunum. The breaking down of food happens mostly in the duodenum. The ileum and jejunum, on the other hand, see to it that the nutrients are absorbed into the bloodstream.

The digestive system empties or moves the food particles from one section to another using a large network of muscles, hormones, and nerves. This part of the process is referred to as "motility."

The walls of the small intestine absorb the nutrients from food and pass them on into the bloodstream. The waste or residue moves on to the large intestine (also called the colon or large bowel).

The pancreas, liver, and gallbladder are accessory organs that play an essential role in the digestion of food in the stomach and small intestine.

Pancreas

The pancreas is responsible for the secretion of digestive enzymes into the small intestine. The enzymes break down carbohydrates, fats, and protein into compounds that the body can use.

They break down carbohydrates into simple sugars. They break down fats into glycerol and fatty acids. They break down protein into amino acids.

Liver

The liver has manifold functions. Within the digestive system, its primary function is to process nutrients. It secretes bile into the small intestine to help in the digestion of fat. It helps to detoxify the potentially damaging chemicals in food and drugs.

Gallbladder

The gallbladder is a pear-shaped organ found under the liver. It functions as a reservoir for the bile that the liver produces. The absorbent lining of the gallbladder concentrates the bile.

Cholecystokinin is a digestive hormone that is released when food enters the small intestine from the stomach. The release of cholecystokinin signals the gallbladder to secrete the bile through the common bile duct into the small intestine.

Large intestine (Colon)

Digestion also results in waste products composed of fluid, undigested food, and older cells that come from the lining of the digestive tract.

The large intestine is a muscular tube connecting the small intestine to the rectum. It consists of the ascending colon, the descending colon, the transverse colon, and the sigmoid colon.

The large intestine receives the waste products from the small intestine. It processes the waste products so that the bowels can be emptied easily.

The large intestine absorbs the water and by doing so solidifies the wastes (stool). It stores the stool and when it becomes full passes the stool into the rectum. When this happens, a person feels the urge to defecate or move the bowels.

Rectum

The rectum is a chamber that joins the long intestine to the anus. It receives the stool from the large intestine and holds it until it is time to evacuate it.

When stool or gas enters the rectum, the brain receives a message from sensors. The brain then signals the body to start the release of the contents. It tells the sphincters to relax and the rectum to contract so that the contents are discharged. If it is not an opportune time to dispose of the contents, the brain tells the sphincters and rectum to act accordingly so that the sensation of wanting to expel stool or gas temporarily passes.

Anus

The anus is found at the end of the digestive tract. It is the opening through which the stool passes to leave the body. It has a muscular ring called the anal sphincter which closes the anus until it is time to move the bowels.

Steps in the Digestive Process

The cell membranes cannot directly access the large molecules that food contains. The food has to be broken down into smaller particles so the body can harness its organic molecules and nutrients.

Digestion or obtaining energy and nutrients from food is a process that involves several steps.

Ingestion

This is the first step in obtaining nutrients from food. It is a process by which food is taken through the mouth and broken down by the teeth and saliva.

Once food goes into the mouth, the teeth, tongue, and saliva help to turn the food into bolus. Bolus refers to the round mass of saliva and food that forms in the mouth when a person chews food.

(In the typical digestive process, the bolus is swallowed and goes down the esophagus to reach the stomach. The bolus combines with gastric juices in the stomach and turns unto chyme. The chyme goes down the intestines and is further digested and absorbed into the blood stream. The undigested part of the chyme or the waste product is eventually discharged as feces).

The process of chewing is called mastication. The food is chewed or masticated into smaller pieces before it is swallowed and further processed by digestive enzymes.

Mastication is one of the primary factors in effective digestion. It is essential for digesting foods like fruits and vegetables that have indigestible cellulose particles that need to be broken down physically.

Mastication is also important for one other reason. Digestive enzymes are only effective on the surfaces of particles of food. They work more efficiently when food is masticated well so that it breaks down to a size that makes it easy to swallow.

While food is masticated, the saliva that the salivary glands secrete helps to chemically process it. Saliva has mucus, the digestive enzymes amylase and lingual lipase, and water. Saliva hydrates the food particles for taste, chemically reduces them, and lubricates them so they can easily be pushed down the length of the esophagus. The combined mechanical and chemical processes help to make swallowing easier.

Drugs and other inedible or harmful substances can be ingested as well. Pathogens like parasites, bacteria, and viruses can be transmitted through ingestion. When this happens, a person becomes susceptible to cholera, polio, hepatitis A, and other similar communicable diseases.

Propulsion

Propulsion refers to the way by which food moves through the digestive tract. This consists of swallowing and peristalsis.

Peristalsis is the primary means of propulsion. It refers to how the muscles lining the walls of the digestive organs alternately contract and relax in order to move food forward along the digestive tract.

Mechanical breakdown

This refers to the process by which food is broken down into smaller particles.

This includes chewing and the churning of the stomach. It includes the segmentation that occurs in the small intestine as the intestinal walls goes through rhythmic muscle contractions that force food to move back and forth and break up, mix with the digestive juices, and ease absorption.

Chemical digestion

This refers to the process of using the enzymes found in the digestive tract, particularly in stomach and small intestines, to chemically break down food into simpler and more absorbable molecules.

Absorption

This refers to the process by which the adjacent lymphatic vessels or blood absorbs the nutrients through the walls of the digestive tract.

Defecation

This refers to the process by which undigested food (waste product) is eliminated as feces through the anus.

Regulatory Mechanisms

Endocrine and neural regulatory mechanisms help the digestive system to function optimally. They stimulate digestion and absorption through chemical and mechanical activities.

Neural Controls

The walls of the digestive tract have nerves that connect the digestive system to the brain and the spinal cord (the central nervous system). These sensors or receptors help to regulate and control certain digestive functions.

These receptors bring in important information. They sense it when the stomach expands because of food. They can tell if the food that is eaten has been adequately broken down. They can tell what type of nutrients (carbohydrates, proteins, and/or lipids) the food contains. They can sense how much liquid there is.

When these receptors are stimulated, they rouse the appropriate reflexes to advance the digestive process. For instance, they may send signals to activate the glands to produce digestive juices. They may send signals to stimulate the muscles of the digestive tract and activate peristalsis or segmentation to push food through the tract.

Hormonal Controls

The digestive process involves certain hormones. These hormones are responsible for communicating to the brain the sensation of fullness or hunger. They send signals to the stomach to secrete digestive juices.

For example, the stomach secretes the hormone gastrin when there is food. The hormone is responsible for the secretion of gastric acid. The small intestine produces the hormone secretin which stimulates the pancreas to secrete bicarbonate. It also produces cholecystokinin which prompts the pancreas to secrete enzymes, the liver to secrete bile, and the gallbladder to release stored bile.

Chapter 15. Metabolism and Human Nutrition

Put simply, human nutrition is the sustenance of the human body with food. Usually, people are focused on how this sustenance, when taken in beyond the limits recommended by daily intake, can store fat. There is much emphasis on how the quality and quantity of food can affect a person's appearance. However, human nutrition is much more than that.

When a person ingests food, many processes occur so that the human body can utilize the nutrients that the food contains. The previous chapter has provided a detailed explanation of how digestion happens. Digestion supports the nourishment of the human body by being especially designed to break food into smaller pieces and basic compositions.

It is difficult to imagine huge pieces of food being automatically absorbed by the body. The body, fortunately, is designed like an engine that is specifically built to absorb nutrients and release the energy from food.

Functions Of Metabolism

The conversion of food to energy is just one of the many roles of metabolism. Metabolism has three main purposes:

1)   Food to energy conversion - This role is vital to the survival of the human body. Without metabolism, the human body will not be able to perform activities that it must be able to do to survive.

2)   Produces building blocks: With the aid of proteins, metabolism helps build tissues in

the body. A good metabolism helps a person heal from illnesses faster.

3)   Nitrogenous Waste Elimination: When protein goes through metabolism, nitrogenous wastes are produced. Examples of such are ammonia, creatinine, and creatinine. After they have been broken down into smaller portions, the nitrogenous wastes are able to leave the body by way of the liver or kidney.

Metabolism and human nutrition are not just closely related but codependent.

Nutrition And Energy Supply

Food is ingested to provide the body with energy. Energy that can be obtained from food is measured in kilocalories. Most people are familiar with the word "Calories".

There's this notion that the standard daily requirement is 2,000 calories. However, the total recommended amount of calories still depends on the person's age, height, and gender. For example, the 2,000 calorie diet is usually what is recommended to an average-sized woman. For men, the recommended daily intake is 2,500 calories.

If energy supply is the sole concern, then the average daily intake should work. The amount of calorie intake can just be adjusted according to different purposes and circumstances. For example, people have to eat less calories than what is recommended to lose weight. They may also opt to exercise, thus using the energy that they have absorbed into their bodies through food.

Nutrition Through The Stages Of Life

Just as gender and weight affects nutrition, the different of stages of life can do the same. Here is a look at human nutrition through some life phases:

From Infancy To Adolescence

Infant nutrition is seemingly simplified through breastfeeding. The infant gets hungry, latches on to its mother, and takes in the needed nutrients. This is complicated when the mother does not produce enough milk. In rare hormonal cases, the mother may not produce milk at all. This is when there is a need to find alternatives, in which cow's milk is the frontrunner.

Childhood presents other threats and complications. A school age child usually uses a lot of his energy on various activities and may refuse to eat fruits and vegetables. With childhood obesity and heart disease on the rise, proper care should be taken when preparing the meals for children. Their age and height should be considerations when determining ideal weight.

With teenagers, proper nutrition is threatened by their busier schedule. There are also many changes that are happening in teenagers' bodies. The girls, especially, can be negatively affected by body image issues. This may tempt them to starve themselves for the sake of achieving society's ideal.

Adulthood And Old Age

Even when the human body reaches its full growth potential, it still needs proper nutrients. Being able to meet the daily required amounts can affect the life expectancy of an individual. It is also important to remember that with muscle mass loss and a general aging of the human body, metabolism slows down.

A younger adult can continue supporting his body through basic daily calorie counting. He can also make sure that his meals are well-balanced. Vegetarians may have to find alternatives to some nutrients, such as protein, to keep that balance.

With an older adult, however, many substantial changes are happening. He may be losing teeth or his taste buds may not be working as they were before. To make things worse: as the body ages, its metabolism slows down. Even if a person does not eat more, he will more likely retain more weight as he grows older.

Pregnancy And Lactation

Pregnancy is a special time when the mother is not just concerned about her nutrition but about her child's, as well. Nutrition is important during this period in a woman's life; underweight and overweight women have a tendency to go through pregnancy complications.

An underweight mother, for example, may have a higher risk of giving birth to a premature or underweight baby. She should be able to gain a little more than 35 pounds during the course of her pregnancy. Anorexic mothers should be provided with more nutritional guidance and may even have to be monitored by a therapist.

On the other hand, an overweight mother has higher risks of diabetes and high blood pressure. These conditions can endanger the baby, especially if they trigger early labor. Less weight gain is recommended for overweight pregnant women.

Generally, pregnant women with no weight issues are recommended to add an extra 500 calories daily to sustain a healthy pregnancy. Alcohol and caffeine consumption, as well as excessive vitamins, are not advised during any pregnancy. Folic acid, in vitamin or natural form, is recommended.

The Effects Of Different Nutrients On The Body

Food provides energy that is needed for a human body to perform activities. A focus on the different nutrients will specifically show how some foods can affect body building, healing, and maintenance.

1.   Carbohydrates

Carbohydrates are mostly known for their energy yield. Among the different types of carbohydrates, glucose is the one that is easily broken down. It is mainly absorbed and use by the brain, central nervous system, and the muscles.

On the other hand, there is fiber. Fiber cannot be digested, but is necessary in preventing constipation and life-threatening conditions such as colon cancer and heart disease.

Carbohydrates are measured according to their glycemic index. Foods with high glycemic index has a fast rate of converting carbohydrates into blood sugar. Examples are white flour, potatoes, and anything else that is refined. Low glycemic foods include whole wheat, whole grain, and high fiber cereals.

Unused carbohydrates may also be converted into fat deposits by the liver.

2.   Proteins

Proteins are responsible for the formation of antibodies, enzymes, and hormones. They are usually known for their body building properties. They are made up of carbon, hydrogen, oxygen, and the inorganic molecule nitrogen. Hemoglobin, which carries oxygen in red blood cells, is a protein.

3.   Lipids

Lipids store energy to regulate temperature, protect organs, and cover nerve cells as myelin. They are found in every cell of the human body and can be very useful. However, there is a need to control the intake of trans fats and saturated fats. One way of doing this is by substituting vegetable oil for canola oil or butter, for example. Note that fats can generate twice as much energy compared to other nutrients. While this may seem like a complete advantage, remember that fats can be stored when unused.

4.   Vitamins and Minerals

These are generally recommended, but there are also acceptable daily limits. Being aware of these limits may help people utilize the optimal potential of the nutrients without abusing them. Sometimes, when food products are able to deliver the daily limits, there is no need to use multivitamins.

Metabolic Physiology

Metabolic physiology is the study of how metabolism functions. As previously discussed, the metabolism of humans can vary. Those with higher level of activities will more likely have higher metabolism. Old age, on the other hand, prompts the slowing down of metabolism.

In popular culture, the topic of metabolism and its physiology are usually discussed in conjunction with weight loss. People are striving to achieve high metabolism because of a desire to lose fat quickly. However, there are those who consider this as a problem because a too lean physique may not exactly project a healthy image.

Types of Metabolic Reactions

There are two main types of metabolic reactions: anabolic and catabolic.

1.   Catabolic – includes processes that require the breaking down of molecules to release energy.

2.   Anabolic – includes processes that require the building up of molecules for energy consumption. This is powered by the previous type.

What Controls the Rate of Metabolism?

The following factors can affect the basal metabolic rate (BMR):

- Thyroxine, a thyroid-produced hormone, determines the rate at which the chemical reactions of metabolism are processed.

- The pancreas releases insulin, which in turn prompts anabolic processes when they are needed (e.g. after a meal).
- The rise in physical activities of a person can trigger a faster metabolic rate.
- The amount of muscle and fat in the body also determines the metabolic rate. More muscle in the body can result to a higher BMR.

Whether one should opt for the fast and slow of it, metabolism is more than just a means to tweak your body fat content. It is also important to maintain a healthy, regulated form of it.

Metabolism in Action

Take, for example, a person who eats a spoonful of sugar or adds it to her tea. This is a good example because half of a person's diet consists of carbohydrates. Sugar belongs to the carbohydrate group.

In the body, the sugar molecules are broken down into smaller, absorbable molecules. This happens together with the release of energy, which is part of the catabolic process.

This process can be represented by the formula shown below:

$$C_6H_{12}O_6 + 6\ O_2 \text{-----}> 6\ CO_2 + 6\ H_2O + energy$$

The energy may be used for the body to function. This newly released energy may be used to warm the body during a cold day or used to power the leg muscles when the person runs. The energy may also be used to maintain the body's functions, which means that the process falls under the anabolic type.

Metabolism physiology analyzes all the different chemical reactions that are going on inside the body during and after it has ingested, broken down, absorbed, and utilized the smaller molecules. Imagine all the processes that a body is capable of that can begin to shut down when a person starves himself. These processes are vital to the maintenance of proper health and well-being.

Chapter 16. Urinary System Fluids, Electrolytes, and the Acid-Base System

The primary **urinary system fluid** is urine. It transports waste products and excess fluids when exiting the body. This is necessary because the body ingests food and liquids several times a day. Some of the foods and liquids get absorbed and utilized by the body. Nevertheless, there are portions that will never be used and should not remain in the body. These are wastes and excess fluids, and one way that they can be expelled is through urine. The urinary system is fundamentally a drainage system for the body.

Overview of the Urinary System

The urinary system is composed of several parts. The main parts are the kidneys, bladder, and urethra. As a system, it is responsible for pH levels and water level regulation. It even affects a person's blood pressure and red blood cells production. Moreover, it regulates the amount of electrolytes (potassium, sulphate) in the body and aids in strengthening the bones.

The kidneys consist of three main parts: the renal cortex, the renal medulla, and the renal pelvis. The renal pelvis squeezes out the urine from the kidneys, through peristalsis movement. The kidneys usually hold 20% of a person's blood volume and performs its role throughout the day.

The Processes That Lead To Urinary System Fluids

Blood enters the kidneys through the renal arteries, which in turn branch out into several pathways. It goes through nephrons, which are the urinary system's main filtration units.

This urinary system is regarded as the cleaner for other processes in the body. The digestive system aids the body in absorbing vital nutrients. Digestion and metabolism, however, can result to some unwanted and possibly dangerous compounds.

Even vital processes can produce waste. For example, the body requires protein for growth. However, protein metabolism produces ammonia, a toxic chemical. The liver assists by converting this to a less toxic chemical, urea. Urea, in turn, is a significant part of urine.

The production of urine, however, happens through three main steps: glomerular filtration, tubular reabsorption, and tubular secretion.

Phase 1

When the blood passes through the glomerulus, the bigger parts – blood cells and proteins – are blocked out. These remain in the blood stream.

Phase 2

The filtrates, as the wastes that managed to pass through are now called, will be passing through the proximal convoluted tubule, nephron loop, and distal convoluted tubule. This long trip provides the blood a chance to reabsorb useful portions of the filtrates. The reabsorption part of the process is important because not all the fluids should be allowed to leave the body or the person will die of dehydration.

Phase 3

At this point, the filtrate is already mostly urine. It will pass the distal convoluted tubule several times. The medulla, which produces salt in Step 2 to draw out water from the filtrates, will be drawing more water from what may be considered urine based on its components. It is in Step 3 that the body tries to reabsorb some of the urea, which may be useful to the kidneys. The urea makes the medulla saltier, as to draw more water to remain in the body. In the end, 120 to 150 quarts have to be filtered to produce what may seem a mere 1 to 2 quarts of urine.

Excess fluids and other substances found in the bloodstream also become part of the urine. These fluids are utilizing urine as their transportation, with the urinary system acting as the pathways.

Urine is the end-product of urinary system fluids. It is stored in the bladder. The bladder fills when it contains about 1 and a half to 2 cups of urine. When a person's bladder is full, he pees and the urine is excreted by way of the urethra. The urine may revert back to its ammonia form, which is what creates the pungent smell.

Other Metabolic Wastes and Foreign Substances in the Urine

Much focus has been given to ammonia, which is a byproduct of protein metabolism and the further breakdown of compounds by the liver. Ammonia is a nitrogenous waste and can be very dangerous.

Having excess ammonia is different from just having excess water. Water retention can be dangerous, as well, but may be remedied with less invasive procedures. High ammonia concentrations mean that the person has a urea cycle disorder, a genetic disorder, or even liver and/or kidney failure. Other protein metabolism wastes include creatinine and uric acid.

Excess urea can wreak havoc in the human body. Adults can experience confusion and fatigue. When not remedied, the condition can result to coma or death. Children cane experience seizures, breathlessness, or even death. There is an emphasis on the word "excess" because the urea does provide some benefits to the kidneys.

Electrolytes

The urinary system is also set to get rid of other metabolic wastes such as electrolytes such as calcium, potassium, phosphates, sodium, and sulphates. Electrolytes are minerals that are used in various essential processes in the body. They are ingested in the form of food and drinks and are absorbed into the body's fluids particularly blood, sweat, and urine. Electrolyte balance also relies on the amount of water the body contains. Excess amounts are also included in the urine to be excreted out of the body.

Electrolytes should not just be viewed as waste, however. They are vital to the proper functioning of the human body's cells and organs. They are responsible for normal muscular contraction, regulated pH levels, hydration, and the conduction of nerve impulses.

Electrolytes carry positive (+) or negative (-) electric charges, hence the name.

The proper filtration and reabsorption processes are vital in keeping the level of electrolytes stable in the body. If the levels are either too high or too low, the body will show some adverse symptoms, some of which can be severe.

Electrolyte imbalance can cause weakness, irregular heartbeats, diarrhea/constipation and more. If not remedied, it can even cause damage to the kidneys. Conversely, kidney disease can cause electrolyte imbalance.

Acid-base Regulation

One of the vital roles of the urinary system is **acid-base regulation**. The urinary system also actively participates in ensuring that the compositions of urine and blood are regulated.

One of the easiest ways to check if the composition of urine is stable would be to observe its appearance. Properly hydrated people will produce clear urine, while dehydrated ones will produce a yellowish or darker type of urine. Clear urine also does not smell as much, but flushing the toilet is still important. When some of the urine reverts to ammonia, there will be a strong smell.

Acids and Bases

The stability of acids and bases is not just quantified by the appearance of urine. Before the **acid-base regulation** of **urinary system fluids** are discussed, it is important to know the pH values of an acid and of a base. A solution with a base above 7.0 pH is a base while a solution with a base below 7.0 pH is an acid.

To fully understand what acids and bases are like, the examples of ammonia solution and lemon juice should be considered. Lemon juice is intrinsically associated with acidity. It does have a pH of 2.0, which is below 7.0. The more acidic substances, such as lemon juice, are more likely to give away hydrogen ions (H^+). On the other hand, an ammonia solution has a pH of 11.0, which is higher than a 7.0. The pH of urine is at 7.0.

Chemical Buffers

The body has a tightly regulated pH balance, with a range of 7.38 to 7.42. Ammonia and bicarbonate, as well as other chemicals, acts as chemical buffers. Chemical buffers ensure that the body stays within the ideal pH range. The acid-base balance is one form of homeostasis. Homeostasis of the human blood composition in particular is vital to survival.

Phosphates, hemoglobin, and other proteins can also act as chemical buffers. Phosphates act as chemical buffers in two forms: sodium dihydrogen phosphate (a weak acid) and sodium monohydrogen phosphate (a weak base). The weak base can dilute a strong acid to produce a weaker acid and salt. On the other hand, the weak acid can dilute a strong base to produce a weaker base and water.

Proteins, such as hemoglobin can also aid the body in its rigid system regulation. Whenever carbon dioxide is converted into bicarbonates, hemoglobin buffers the hydrogen ions that have been released. This helps regulate the blood's pH.

Balance and Imbalance

When the body contracts the disease, the blood can be compromised. This can cause an imbalance. The blood cannot perform as expected. The same can be said if a person develops a kidney disease.

When homeostasis is disrupted, proteins change its form and may be digested. A disturbance of the acid-base system can also result to the disfunction of enzymes. If acid-base imbalance is further ignored, it may eventually lead to death.

While the kidneys can have difficulties readjusting to and coping with a breakdown in the balance, compared to the lungs, overall renal physiology has some backup. For example, the tubular cells reabsorb extra bicarbonates when acidosis occurs.

Acidosis is an imbalance that leans toward a greater level of acid than required. The pH level will have dropped lower than 7.0, or even just below 7.38. Acidosis can be caused by aspirin or methanol poisoning, dehydration, or kidney disease.

Conversely, the kidneys will excrete extra bicarbonates in the case of alkalosis. Alkalosis happens when the pH goes above 7.42. This means that the blood composition is more alkaline or base than it is supposed to be. Chloride use due to vomiting, aspirin poisoning, and kidney disease may cause alkalosis. When the pH goes beyond 7.65, the sick person has a higher possibility of dying.

The acid-base regulation of the body is not just reliant on the urinary system. However, the system has a vital role in keeping things balanced. It does have the capacity to reabsorb fluids or to excrete wastes. It is also in urine testing that an imbalance can be quickly and primarily detected.

To summarize, the kidneys roles in acid-base regulation are 1) to reabsorb bicarbonate from filtrates or urine and 2) to dispose of hydrogen ions through urine. Kidneys continuously work to ensure that a balance is reached or maintained.

The urinary system is a vital player in keeping the body at a state of homeostasis. It absorbs nutrients that can be used to build and maintain the body's systems and functions. On the other hand, it expels waste byproducts to avoid toxicity in the bloodstream through an important but simple form of waste – the urine.

Chapter 17. Reproductive System Physiology

The human reproductive system is a cluster of organs within the body of a man and a woman, with the ultimate purpose of creating offspring.

Another purpose of the human reproductive system is to give pleasure and form a bond between a man and a woman through sexual intercourse.

The Reproductive Process

Human reproduction starts with fertilization through sexual intercourse. The man's erect penis is inserted into the woman's vagina until ejaculation of the sperm cell into the vagina.

The sperm travels from the vagina to the cervix and settles in the uterus for possible fertilization of an ovum.

Upon fertilization, fetus gestation occurs in the female uterus for approximately nine months until labor occurs.

During labor, the female uterine muscles contract and the cervix dilates until the baby passes out through the vagina giving birth to a new offspring.

Male Reproductive System

At the start of puberty, the hypothalamus begins secretion of the gonadotropin releasing hormones or GnRH.

In reaction to the GnRH, the pituitary gland releases FSH or follicle stimulating hormone and LH or luteinizing hormone into the male system.

The FSH flows to the testes to stimulate the Sertoli cells. This helps nourish the sperm cells produced by the testes and activates the spermatogenesis process.

LH also flows inside the testes to stimulate the interstitial cell or Leydig cells to produce and release testosterone to the testes and the blood.

During intercourse, vasodilation happens as a result of physical or psychological stimulation. This increases the blood flow to the penis. It becomes enlarged and firm, the skin in the scrotum tightens, and the testes pulled up the body.

During peak of coitus, muscles in the epididymis, prostate gland, seminal vesicles and the vas deferens contract.

Simultaneously, a sphincter muscle located at the base of the bladder contracts to prevent urine from leaking into the urethra. A second sphincter also contract to close the entrance of the urethra into the penis.

This will force the sperm and the semen into the urethra. This is the start of orgasm and once started the man will ejaculate and orgasm fully. If stimulations stops before orgasm, the contraction and physical effect on the penis will subside.

Repeated and prolonged stimulation without orgasm can lead to discomfort for the man. This discomfort is what some refer to as "blue balls".

During the expulsion stage or the release of the sperm from the urethra, the second sphincter relaxes to allow semen to enter the penis. Simultaneously, strong muscle contractions occurs around the base of the penis to expel the semen from the body.

After ejaculation, a recovery phase period follows, characterized by the feeling of relaxation associated with oxytocin and prolactin neurohormones.

The Male Reproductive Organ

The male reproductive organ includes external and internal organs and structures that aids in the reproductive process.

Male external genital structure

Penis

The Penis is an organ resembling a pendulum hanging from the sides and front area of the pubic arch. It is made up of spongy tissues that makes it turgid and erect when filled with blood.

It consist of the glans, the shaft and the root.

The glans is the head of the penis that contains nerve endings and its rim is called the Corona. Covering the glans is the foreskin or prepuce.

The shaft is the pendulous length from the body to the glans.

The root is the part of the penis that connects the shaft into the pelvic cavity.

penis is made up of spongy tissues that makes it turgid and erect when filled with blood.

The *corpus spongiosum,* which is the mass of tissue surrounding the urethra where blood flows when penis is erect.

The corpus cavernosa, which is a pair of spongy tissue where blood is contained during erection.

The urethra is a tube inside the penis where semen travels during ejaculation.

Scrotum

The scrotum is the pouch of skin located in the part of the abdominal wall. The colloquial term for the scrotum is "balls".

The scrotum keeps the testis cooler than the body temperature.

The surface of the scrotum is a ridge it into two compartments. Each compartment contains one epididymis and one of each two testes.

The ridge extends to the under surface of the penis up to the back, in the perineum line to the anus.

The scrotum has four layers.

The *skin,* which is outermost layer. It is wrinkled in appearance, thin and pigmented.

The *superficial fascia,* which is the first inner layer. This is a membranous and fatty layer and contains the sympathetic nerve fibers and the reason why the skin for the wrinkled skin.

The *spermatic fasciae,* which is beneath the *superficial fascia.* It consists of three layers originating from the three layers of the anterior abdominal walls.

The external spermatic fascia, which originated from the aponeurosis of the external oblique muscles; the cremasteric fascia originating from the external oblique muscles; and the internal spermatic fascia from the fascia transversalis.

The last layer of the scrotum is the *Tunica Vaginalis,* which is within the spermatic fasciae. It covers the medial, anterior, and lateral surfaces of the two testes.

Male Internal genital structure

Testis (plural Testes)

There are two testes. These two oval-shape male organ produces the sperm and testosterone. The testes develop inside the abdomen and move to the scrotum before birth.

The Tunica vaginalis and the tunica albuginea covers is the cover protection of the testes. Inside the surface, there are fibrous septa that the interior into lobules. Within each lobule are coiled seminiferous tubules, this opens into structured channels called rete testis.

The rete testis is connected to the upper end of the epididymis through small ducts.

Epididymis

The epididymis is a long, lone-coiled duct that passes along the posterolateral side of the testes. It is covered by the tunica vaginalis. The epididymis is the storage and maturation holding area of the sperm.

The two components of the epididymis are the efferent ductules, which forms the head of the epididymis and true epididymis, which forms the body and tail of the epididymis.

Vas deferens (Ductus Deferens)

Ductus deferens or vas deferens is the Latin word meaning "carrying-away vessel".

The vas deferens is thin tube that transports sperm from the epididymis to the urethra in preparation for ejaculation.

The vas deferens dilates to form the ampulla and goes to the ejaculatory duct. The ejaculatory duct goes through the prostate gland and connecting with the urethra.

Accessory Glands

The penis has three accessory glands that produces fluids to nourish the sperms.

The *seminal vesicles* are structures attached to the ductus deferens near the bladder. The *seminal vesicles* produces a yellowish, sticky flued that contains the fructose.

The *prostate gland* is the backdrop of the ejaculatory duct from the base of the urethra, below the bladder. The *prostate* produces a combination of sperm cells, seminal fluid, and prostate fluid called the semen.

The *bulbourethral gland* are the two glands found on the side of the urethra below the prostate. The *bulbourethral gland* produces a slippery, clear fluid that goes directly to the urethra.

Spermatogenesis

The main function of the male reproductive system is to produce testosterone to maintain the reproductive function and start the spermatogenesis process.

The testes is the conduit responsible for testosterone and sperm production and transport to the female reproductive system.

Spermatogenesis is the formation of the sperm cells. It starts to occur at puberty and each cycle begins every 13 days.

There are three stages of spermatogenesis.

The first stage is mitosis. Diploid cells from the outer walls of the tubules constantly multiply each day by 3 million primary spermatocytes.

The primary spermatocytes moves to the ad luminal pockets of the seminiferous tubules to duplicate its DNA and produce two haploid secondary spermatocytes through the meiosis I process.

The secondary spermatocytes goes straight to meiosis II process to produce haploid spermatids. The spermatids developed a tail, midpiece and a head to form the sperm or spermatozoa.

Maturation of the sperm happens through the influence of the testosterone. Sperm is stored in the epididymis until ejaculation.

Female Reproductive System

The female reproductive system serves several functions.

It produces the female egg cells called the ova or oocytes needed for reproduction. Other functions include conception, breastfeeding, menstruation, and production of the sexual hormones.

Menstruation signals the bodily changes transforming the female body. Menstruation starts on the average at the age of 12.8.

Blood color varies from red to dark maroon and the amount of bleeding differs for every woman.

The menstrual cycle occurs average of 28 days and the women may experience cramps, mood swings, fluid retention, weight gain, diarrhea, breast tenderness and constipation.

The menstrual cycle begins with maturity of the egg cells through oogenesis process and follicle development.

Oogenesis begins with the development of oogonia thru the change of the primary follicles into primary oocytes. This change is the process of Oocytogenesis.

This process becomes complete either before or after birth in humans. Primary oocytes complete maturation during the menstrual cycle through meiotic divisions.

FSH, estrogen, LH and progesterone helps bring for the first meiotic division. As the FSH and the LH increases, this stimulates ovulation of the oocyte in the fallopian tubes.

Once the follicle matures, the ovum is shed and the corpus luteum develops. This is what is referred to as the ovarian cycle.

During the luteal phase, the corpus luteum forms and the secretes hormones, particularly progesterone, preparing the endometrium of the uterus ready for implantation.

If implantation happens, the corpus luteum continues, but if there is no implantation, the corpus luteum will slow down resulting to menstruation.

The Female Reproductive Organs

The female reproductive organs may be divided into external structures, internal structures and accessory organs.

The female reproductive organs are related to copulation, fetus fertilization, growth, development and birth.

<u>The female external genital structure or the Vulva</u>

The area containing the external genital structure of a woman is called the vulva. The vulva includes the mons pubis, labia majora and minora, clitoris, urethral opening, vaginal opening and the perineum.

Mons Pubis

The mons, as it is often called, is the triangular fatty tissue covering the pubic bone protecting the pubic symphysis. Hormonal development during adolescence triggers growth of pubic hair on the mons pubis. Hair varies in color, amount, texture, curliness and thickness.

Labia Majora

The labia majora is the outer lips of the urethral opening. The labia majora is darker in pigmentation and covered with hair and sebaceous glands. It protects the urethral and introitus opening and contains sweat and oil-secreting glands.

The labia majora turns flaccid after childbirth and with age.

Labia Minora

This is often called "inner lips".

It is made up of erectile, connective tissue. It darkens and swell during copulation.

The labia minora is right inside the labia majora.

It is more sensitive and responsive to stimulation and the labia minora contracts during intercourse.

Clitoris

The clitoris is a highly sensitive mass of nerves, erectile tissue and blood vessels. It is made up of a gland and a shaft that swells with blood during intercourse. It can be stimulated such that the woman can achieve orgasm.

Urethral Opening

The urethral opening is located below the clitoris. It is the tubular transport for urine discharge from the bladder. The female urethra is shorter than the male urethra.

Vaginal Opening

The vaginal opening is located below the urethra. The vaginal opening is often covered with a thin lining called the hymen.

The vaginal opening is where menstrual fluids are discharged.

Presence of the hymen cannot be use as reference to determine virginity. There are women born without hymens and there are many reasons outside of sexual contact where hymens are perforated.

Perineum

The perineum is the tissue and muscle found between the vaginal opening and the anal canal.

The perineum contains abundant nerve endings making it sensitive to touch. It supports the lower parts of the digestive and urinary tract.

During childbirth, an episiotomy is performed to widen the vagina opening.

Female internal genital structure

The internal genital structure of the woman consists of vagina, cervix, uterus, fallopian tubes, and the ovaries.

Vagina

The vagina is located between the rectum and the bladder. It connects the cervix to the external genitals.

The vagina is the tunnel for the menstrual flow. It allows uterine secretions to pass down through the vaginal opening.

It also serves as the birth canal during childbirth and produces lubrication during intercourse with the help of two Bartholin's gland.

Cervix

The cervix connects the vagina to the uterus. The cervical opening to the vagina is small but it dilates during childbirth to allow passage of the fetus.

The small cervical opening also serves as protection against intrusion of foreign bodies entering the uterus.

Uterus

The uterus or the womb is a pear-shaped organ about the same size of a clenched fist.

The uterus is made up of the myometrium, endometrium, and the perimetrium. It also comprises blood-enriched tissue that is released during the monthly menstrual cycle.

The uterus is flexible enough to expand during pregnancy and its powerful muscles pushes the fetus through the birth canal.

Round ligaments, broad ligaments, cardinal ligament, and utero-sacral ligament support the uterus.

Fallopian Tubes

Fallopian tubes or uterine tubes serves as a pathway for the ovum. This is where fertilization of the egg by the sperm cell happens.

Fertilized eggs travels approximately 6 to 10 days through the fallopian tube where it will be implanted.

Ovaries

The ovaries are the female sex glands or gonads. The follicles in the ovaries produces female sex hormones progesterone and estrogen. These hormones are responsible for preparing the uterus for arrival of the fertilized egg.

A woman normally has 400,000 follicles or immature eggs and during her lifetime can release 400 to 500 follicles.

Other female structures

Breasts contains mammary glands that produces milk after giving birth. This is part of the after birth care to complete the reproductive cycle.

Sucking of the nipple stimulates the pituitary gland to release prolactin and oxytocin.

Chapter 18. Developmental Anatomy and Physiology

The term developmental anatomy or embryology refers to the study of differentiation and growth of the organism from cell to birth.

Human Development

The human development begins with fertilization – a process in which the egg cell is penetrated by the sperm cell.

The fertilized ovum or zygote will undergo the processes of mitosis and cell differentiation. The resulting embryo will be implanted in the uterus – an important female reproductive organ. The embryo will continue to grow and develop inside the uterus until birth.

The Sperm Cell

Sperm cells are gametes or sex cells that are stored in male gonads (testes). These cells contain 23 chromosomes that resulted from meiosis, a special type of cell division.

Sperm cells move out from the epididymis through the vas deferens. These gametes are lubricated by the prostate and bulbourethral glands to form semen. This is a male reproductive fluid that contains millions of sperm cells. The semen leaves the male body through the penis.

The three major parts of the sperm cell are:

The Head

The smooth, oval-shaped head of the sperm resembles the shape of a duck egg. It measures 2.5-3.0 um in diameter and 4.0-5.2 um in length. The head of the sperm cell contains the nucleus that houses the DNA of the cell. The head also stores the enzymes that help break and penetrate the membrane of the egg cell.

The Body (Midpiece)

The midpiece is a slim section of the sperm that has the same length as the head. It contains millions of mitochondria, the powerhouse of the cell. These are the organelles that produce energy in cells. The shape of the body also enables the sperm to move faster.

The Tail

The tail of the sperm, or the flagellum, is a thin bundle of filaments that enables the sperm to move fast. It has a length of 50um and a thickness of 1.25um. The process of capacitation activates the tail movement. In this process, the sperm undergoes cellular changes in preparation for fertilization.

During ejaculation and/or sexual intercourse, a man releases almost 100 million sperm cells from the penis. Only one sperm will fertilize the egg cell of the female. Those that will not reach the egg cell will eventually die.

The Egg Cell

The egg cell, or the ovum, is the female gamete (sex cell) that can be fertilized by a healthy sperm cell. It is considered as the biggest cell in a female body that is not capable of active movement. An egg cell is 30 times wider than the sperm cell and can be seen by the naked eye. A human female can have 7 million ova in her ovaries after birth. These will be released from her ovaries during the menstrual cycle.

The following are the parts of the egg cell:

<u>Nucleus</u>

The heart of the egg cell, or the nucleus, contains the 23 chromosomes or the genetic materials of the female. These chromosomes determine the inherited characteristics of the baby.

Cytoplasm

This is a gelatinous substance that keeps the cell's internal structures (organelles) in place. The cytoplasm contains the mitochondria that supply energy to the egg cell and its reproductive processes.

Zona Pellucida

It is also called the egg wall, or the outer membrane of the egg cell. It is made of glycoprotein layer that helps the sperm cells enter easily. The zona pellucida protects the immature oocytes and embryos in the developmental period. This also provides an efficient communication channel between oocytes and follicle cells during oogenesis. This is the process of producing a healthy ovum. The zona pellucida also regulates the egg-sperm interaction during and after fertilization.

The Corona Radiata

The corona radiata is the innermost layer of the egg cell that is attached to the zona pellucida. The main function of corona radiate is to provide a continuous supply of vital proteins to the egg cell.

The Prenatal Development Processes

Prenatal or antenatal development includes a series of processes that a developing embryo/fetus undergo before birth.

The first two weeks following fertilization are referred to as the germinal stage. Next is the embryonic stage that happens from 3rd week to 8th week. The fetal stage happens from 9th week to birth. The fetal development period is a period of progressive growth of a fetus inside the mother's uterus to its delivery.

The Germinal Stage

Fertilization happens when a small, asymmetric, and motile sperm cell successfully enters the membrane of a large and non-motile egg cell. The egg and sperm chromosomes merge to form a single-celled zygote. Then the germinal stage of embryogenesis follows. The germinal stage is a 10-day period that starts from fertilization to implantation. It begins with the egg-sperm cell union in any of the two fallopian tubes.

The process of cell division happens after 2-3 days following conception. The zygote or the fertilized cell then moves to the uterus in about 4-7 days.

Cell division processes happen in the uterus. The process of mitosis initially divides the zygote into two cells. Then two halves will be divided into four, eight, sixteen, and so on. This cell-multiplication process separates the cells into two masses – the outer cells and the inner cells. The outer cells will form the placenta while the inner cells will give rise to the placenta.

The cell division process will continue and will develop into a blastocyst. A blastocyst is a thin-walled structure that contains a cluster of cells from which the embryo arises. The blastocyst has three important layers:

The Ectoderm, which will form and develop the skin and nervous system

The Endoderm, which will form and develop the respiratory and digestive systems

The Mesoderm, which will form and develop the skeletal and muscular systems

When the blastocyst arrives at the uterus, it attached itself to the uterine wall. This is a process called implantation. The blood vessels and membranes of the uterine wall will provide nourishment for the developing embryo. Implantation is not always 100% successful and accurate process. Successful implantation will temporarily stop a woman's regular menstrual cycle and will cause evident body changes. This is the start of pregnancy.

The Embryonic Stage

At this stage, the mass of cells is called the embryo. The embryonic period starts at the beginning of the 3rd week following conception. This is the time when the mass of cells resembles a human figure. Embryonic development plays a significant role in the brain development of the growing embryo.

Embryonic development has four stages. These are:

1. ***The Morula Stage***

It is an early-stage embryo with 16 blastomeres in a spherical solid within the zona pellucida. If untouched, it will develop into a blastocyst or a hollow ball of cells.

2. ***The Blastula Stage***

The blastula is a hollow sphere of blastomeres (cells) that surround the blastocoele – a fluid-filled cavity. In humans and other mammals, blastula is also called blastocyst. The blastocyst stores the embryoblast, the inner cell mass that forms the fetal structures. It also contains the trophoblast, that is responsible for the formation of extra-embryonic tissues.

3. ***The Gastrula Stage***

Gastrulation is an early embryonic development process that turns the single-layered blastula into multi-layered gastrula. Triploblastic organisms are trilaminar or three-layered. These layers are the ectoderm, mesoderm, and endoderm. They are responsible for the development of different tissues and organs of the embryo. The ectoderm gives rise to the nervous system, neural crest, and epidermis. The endoderm develops the digestive system, respiratory system, and the epithelium. The mesoderm gives rise to bones, muscles, and connective tissues.

4. *The Neurula Stage*

The neurula stage follows after gastrulation. Neurulation initiates the process of organogenesis – the production and development of the internal organs of an organism. The thickened neural plate that rose from the ectoderm will continue to broaden and the linings will start to fold upwards. These upward folds are referred to as neural folds. This phenomenon happens during the 16th week of pregnancy. The folds will continue to increase height and will meet together at the neural crest. In the process of somitogenesis, the somites will form the syndetome, sclerotome, myotome, and dermatome. These somitomeres form the bone, tendons, cartilage, muscle, and dermis (skin). The urogenital tract will be formed by the intermediate mesoderm while the caudal part will be formed by the lateral mesoderm.

The Fetal Stage

The fetal period starts at the beginning of the 36th week and ends at the time of fetal delivery. This period marks more significant changes in brain development. The immature systems and structures that took place in the embryonic stage continue to develop in the fetal period. The neural tubes begin to develop into the brain, spinal cord, and neurons. The neurons will start to migrate into their respective locations. The synapses – connections between neurons, will also begin to develop.

Development of Organs and Organ Systems

1. *Blood*

The haemotopoietic stem cells that form all the blood cells develop from the mesoderm. Blood formation happens in the blood islands, the clusters of blood cells in the yolk sac. Hemangioblasts form the haematopoietic stem cells in the center of the blood islands. These stem cells are the precursor to all types of blood cells.

2. *The Cardiovascular System*

The heart starts to develop, beat, and pump blood on the 3rd week of gestation. The blood islands and cardiac myoblasts in the splanchnopleuric mesenchyme on both sides of the neural plate form the cardiogenic region of the fetus. At day 21, the two endocardial tubes fuse to form a single, primitive tube – the tubular heart.

At this time, the endocardial tubes and vasculogenesis, or the development of the circulatory system, are also starting to develop. This happens on day 18 when the cells in the splanchnopleuric mesoderm (angioblasts) form flat endothelial cells. These cells form small vesicles (angiocysts) that merge to form angioblastic cords. These cords form a pervasive network of plexuses. This network expands through sprouting and additional budding of new vessels. This expansion is called angiogenesis.

3.   *The Digestive System*

The fetal digestive system starts to form on the 3rd week. At 12th week, the main and accessory organs of the digestive system are already in their respective places.

4.   *The Respiratory System*

The lung bud found in the ventral wall of the foregut forms the respiratory system. It happens on the 4th week of gestation or fetal development. The lung bud that forms the trachea and the bronchial buds expand on the 5th week to form the main bronchi. The left and right bronchi will eventually form secondary and tertiary bronchi after the 5th week of gestation. The internal lining of the larynx originates from the lung bud while its muscles and cartilages originate from the 4th and 6th pharyngeal arches.

5.   *The Urinary System*

The three kidney systems – pronephros, mesonephros, and metanephros arise from the intermediate mesoderm. The metanephros system forms the permanent kidney of the fetus. This starts to appear in the 5th week of fetal development. The ureteric bud, a mesonephric duct outgrowth, penetrates metanephric tissues to form the renal calyces, pelvis, and pyramids. Also, the ureter is also formed. Between the 4th-7th weeks of development, the cloaca is divided by the urorectal septum into urogenital sinus and anal canal. The upper region of the uro sinus becomes the bladder while the lower region forms the urethra.

6.   *The Integumentary System*

The ectoderm forms the outer superficial layer of skin or the epidermis. On the other hand, the mesenchyme forms the deeper layer of the skin or the dermis. The formation of the epidermis happens between the 2nd-4th months of development. The ectoderm will be divided to form the periderm – a flat layer of surface cells.

Different layers of the epidermis will be formed through periderm division.

7.   *The Nervous System*

In the 4th week, cranial neural crest cells move to the pharyngeal arches where they develop into neurons. Neurons are formed through the process of neurogenesis.

Development of Physical Features

Face and Neck - develop between 3rd – 8th weeks of gestation

Ears - develop between 6th – 8th weeks of gestation

Eyes - develop between 3rd – 10th weeks of gestation

Limbs - start to develop at the end of the 14th week of gestation

The baby is delivered on the 40th week or around 280 days from the first day of the last menstrual period of the mother. This also marks the end of the fetal stage.

Chapter 19. Immune System Physiology

The immune system is a framework of organs, tissues, and cells. They work together to safeguard the body from foreign invaders, namely fungi, bacteria, and viruses. Since the human body is an ideal habitat for microbes, many microorganisms try to break in daily.

It's this organ system's job to seek, destroy, and keep pathogens out. However, when lymphocytes, the cells of the adaptive immune system, are crippled or attacks the wrong target, autoimmune diseases can develop such as arthritis and allergy.

The immune system is complex. With memory b-cells, it can remember and recognize different microorganisms. With the lymphocytes, it can produce toxins and antibodies that can wipe out microbes that enter the body.

Collectively, lymphocytes are a subtype of white blood cells. According to Mayo Clinic, they defend the body against diseases and the organisms that cause them. There are three types of lymphocytes: natural killer cells (NK cells), B-cells, and T-cells.

Once the right cells receive the alarm, leukocytes—white blood cells—undergo changes. Some subtypes of WBCs produce powerful chemicals toxic to viruses and bad bacteria, whereas some leukocytes regulate the behavior and growth of other immune cells.

In this chapter, you will learn how the immune system works and the functions of the different types of white blood cells.

The Structure of the Immune System

The structure of this organ system is similar to that of the endocrine system. The lymphoid organs are located throughout the body. These organs are home to lymphocytes.

Bone marrow, the tissue at the center of large bones, produces blood cells, including WBCs. Blood cells start as immature cells referred to as stem cells.

There are two types of bone marrow: yellow marrow and red marrow (myeloid tissue). WBCs, RBCs, and platelets arise in myeloid tissues. Some leukocytes, however, develop in yellow marrows.

T-cells mature in the thymus. The thymus is the primary lymphoid organ of the immune system. T-cells are important to the adaptive immune system. T-cells, along with b-cells and macrophages, aids the body's immune response. The body's immune response enables it to defend itself against pathogens and cancerous cells.

The immune system safeguards the body from harmful substances and foreign invaders by responding to antigens.

Antigens are proteins on the surface of bacteria, cells, fungi, and viruses. Nonliving substances, namely drugs, chemicals, and toxins, can also be considered as antigens. Antigens are responsible for triggering an immune response.

Body cells also contain antigens, but specific immune cells determine these substances as "harmless."

The Lymphatic System

The lymphatic system is an integral part of the immune system. It's a network of lymphatic vessels. These vessels transport lymph, the fluid that flows through the system, towards the heart. Fluids and cells are exchanged between the body's arteries and lymphatic vessels. This enables the immune system to monitor invading microbes.

Lymphocytes travel all over the body through the blood vessels and the lymphatic vessels. The structure of the lymphatic vessels is similar to that of the body's arteries and veins.

Lymph nodes are laced along the vessels. Clusters of lymph nodes are located in the groin, abdomen, armpits, and neck. Each node has specialized compartments. This is where immune cells are stored and where they may encounter antigens.

Foreign particles and immune cells enter the nodes through incoming lymphatic vessels. Lymphocytes exit through outgoing lymphatic vessels. They are then transported throughout the body.

The main role of lymphocytes is to patrol the body. They look for foreign antigens everywhere. They gradually return into the lymph nodes to continue the cycle.

Lumps of lymphoid tissues are located in various parts of the body. These tissues are found in the digestive tract, adenoids, appendix, tonsils, and lungs.

The spleen is also a part of the immune system. This organ synthesizes antibodies in its lymphoid tissues. It also removes blood cells and bacteria coated with antibodies through blood and lymph node circulation.

A study that was published in 2009 revealed that the spleen's red pulp is like a reservoir. It contains more than 50% of the body's monocytes.

Monocytes are the largest white blood cells. Upon reaching injured tissues, monocytes differentiate into macrophages and dendritic cells. They promote tissue healing. When a monocyte changes into another cell type, it undergoes cellular differentiation.

The Different Types of Immune Cells and Their Functions

This organ system is responsible for regulating and producing numerous blood cells. These cells are called immune cells. Phagocytes and lymphocytes are examples of immune cells. Each type of white blood cell has a specific function.

Immune cells cooperate with other WBCs to work efficiently. In some instances, they communicate by direct contact or by releasing chemical substances. Further, some cells are designed to store genetic materials of pathogens after primary infection. This helps in fighting re-infection.

As stated, immune cells start off as stem cells in marrows. They grow into cell types, namely phagocytes, B cells, and T cells, by responding to signals and cytokines.

B Lymphocytes

B cells secrete immunoglobulins into the bloodstream. Immunoglobulins or antibodies neutralize pathogens, namely viruses and bad bacteria. They recognize the antigens of pathogens by way of fragment antigen-binding.

Immunoglobulins ambush antigens that are circulating in the bloodstream. These Y-shaped proteins cannot penetrate cells. The T-cells are the ones responsible for attacking virus-infected and cancerous cells.

Each b lymphocyte is programmed to produce one specific immunoglobulin. For example, a b-cell will produce an antibody that can only neutralize a virus that causes influenza, while another will make an antibody that attacks the pneumonia-causing bacterium.

When a b cell encounters a triggering antigen, the cell produces plasma cells. The plasma cells that come from a particular b lymphocyte manufactures a myriad of identical antibodies.

The relation between an antigen and the

antibody made for it is similar to how a key matches a padlock. Whenever an antibody and an antigen interlock, the antibody marks the antigen. T cells will now destroy that antigen.

T Lymphocytes

Unlike b lymphocytes, T cells don't recognize circulating antigens. Instead, their outer covering contains antibody receptors. These receptors are able to detect antigen fragments on cancerous or infected cells.

There are two types of T cells. Th cells or helper T cells regulate and direct immune responses by signaling other WBCs. Some Th cells stimulate B lymphocytes to manufacture and secrete antibodies, whereas others call phagocytes to gobble foreign bodies.

Killer T cells, CTLs, or cytotoxic T lymphocytes perform various functions. They attack infected cells, viruses, and abnormal molecules. CTLs are efficient in getting rid of viruses hiding in body cells. They detect infected cells, as well as foreign invaders.

NK Cells

Like T cells, natural killer cells are lethal WBCs. While T lymphocytes recognize invaders and infected cells by detecting antigen fragments, NK cells attack bodies that lack MHC molecules.

Therefore, NK cells attack many kinds of foreign bodies. Both the T and NK cells slay invaders that have been marked by antibodies. They bind to their targets and kill them with lethal chemicals.

Phagocytes and Other Similar WBCs

Phagocytes are larger than T cells and B cells. They engulf and digest foreign particles and microbes. Monocytes, the largest WBCs, are phagocytes. They circulate throughout the body.

When monocytes move to tissues, they gradually become macrophages. Macrophages use phagocytosis to kill or digest microbes, harmful substances, and cancer cells.

During phagocytosis, phagocytes utilize their plasma membrane to engulf particles or foreign bodies. After this, a phagosome is formed.

A phagosome is a vacuole or an internal compartment. It contains a phagocytosed particle that is enclosed in a white blood cell's cytoplasm.

Special macrophages can be found in organs of the body such as the liver, brain, kidneys, and lungs. These large WBCs have many functions.

They act as scavengers, ridding the body of debris and old and dead cells.

In addition, macrophages display fragments of antigens to attract matching lymphocytes. They also produce chemical signals, namely monokines, vital to immune responses.

The granulocytes, another type of white blood cells and a phagocyte, utilizes packaged lethal substances to disintegrate ingested microbes. The basophils and eosinophils use degranulation to kill microbes nearby.

Degranulation is a cellular process. Specific types of WBCs release cytotoxic molecules from granules. It is utilized by granulocytes and lymphocytes, namely eosinophils, basophils, neutrophils, cytotoxic T cells, and NK cells.

The basophil has a twin—the mast cell. Basophils and mast cells are involved in a variety of inflammatory and allergic diseases.

These cells produce high amounts of histamine and other related substances.

Histamine is a compound that is released by some immune cells as a response to injury or foreign particles, such as dust and pollen. It causes muscle contraction and capillary dilation.

Cytokines

Cytokines are vital components in cell signaling. They aid in governing basic cell activities and coordinating all cell actions. Cytokines enable and bolster immunity, homeostasis, development, and tissue repair. Errors in cell signaling can lead to diabetes, autoimmunity, and cancer.

Interferons, interleukins, and growth factors are cytokines. By exchanging cytokines, the components of the immune system can communicate with each other. They also coordinate the body's immune response.

Some cytokines act as chemical switches. They can turn specific immune cells on and off. The interleukin 2 (IL-2), for example, triggers the system to secrete T cells.

The chemokines, cytokines that attract specific white blood cells, aids in regulating immune responses. A broad range of cells produces cytokines. This includes immune cells, such as T cells, mast cells, b cells, and macrophages.

Chemokines are released at sites of infection or injury. They call immune cells to repair damaged regions or destroy invaders.

The Complement System

The complement system is part of the immune system. This system is composed of twenty-five proteins working together to assist antibodies. They aid in destroying bad bacteria that enter the body.

The proteins also get rid of antigens coated with antibodies. They circulate throughout the body as inactive organic compounds. When an antibody-coated antigen activates the 1st protein in the series, a domino effect will be set in motion.

This process is called the complement fixation cascade. The final step in this process is the stimulation of phagocytes to clear damaged and foreign material. This, however, also causes inflammation by attracting many phagocytes.

The Body's Immune Response

Infections cause various diseases ranging from debilitating conditions, such as chronic hepatitis and pneumonia, to fatal diseases, namely cancer, and AIDS.

If pathogens penetrate the skin, they will pass through the urogenital passageways or the walls of the respiratory or digestive system. Epithelial cells line the passageways. This lining is covered with mucous. Mucosal surfaces inside the body secrete immunoglobulin A (IgA).

Under the second line of defense, T cells, B cells, and macrophages are lying in wait for foreign invaders. Immune cells that do not regard antigen markers are in the bloodstream.

The complement proteins, phagocytes, and phagocytes are circulated throughout the body. They're constantly looking for harmful debris and bad bacteria.

If microbes are able to penetrate the body's barriers, then they will face lymphocytes designed just for their destruction. B cells produce antibodies that mark pathogens and harmful particles. They command T cells to attack antibody-coated antigens. The phagocytes use phagocytosis to engulf and destroy bacteria and debris.

Chapter 20. The Kidney

The kidneys are two bean shaped organs located on either side of the spine. They sit below the ribs and behind the belly. An adult human's kidney can measure 4.3 inches in length. This is equal to the size of a fist.

The main function of the kidney is to filter out waste products from circulating blood. The kidneys regulate circulating blood volume, extracellular fluids, and blood pressure.

Kidneys also regulate sodium, potassium, chlorine and other electrolytes. They are also responsible for the regulation of homeostatic ph. The kidney reabsorbs glucose and is able to produce it as well. Kidneys also produce the hormones renin and erythropoietin.

Nephrons

A Nephron is the functional unit of the kidney. Healthy adults have about 1.5 million of these filtering units per kidney. Nephrons filtrate, reabsorb, and secrete substances to make urine.

A renal corpuscle and a renal tubule comprise a Nephron. The renal corpuscle is the component of the Nephron which filters blood. It consists of a Glomerulus and the Bowman's capsule.

The Glomerulus is a network of filtering small blood vessels. The Bowman's capsule is a cup like sack that surrounds the Glomerulus. The Glomerulus contains filtered fluid from blood.

This tubular fluid collects into the Bowman's capsule. This fluid, now called glomerular filtrate moves to the renal tubule. The filtrate passes through the renal tubule for further processing to form urine.

The Filtration process

Filtration takes place in the renal corpuscle. The renal artery brings blood into the kidneys. Blood goes through the renal artery then into the afferent arterioles.

The afferent arterioles enter and wrap around the Bowman's capsule. It branches to the capillaries of the glomerulus. The capillaries later converge and exit through the efferent arterioles.

The volume of plasma entering the afferent arteriole is at a hundred percent. Only twenty percent of this volume gets filtered in to the Nephron. This percentage is also known as

Filtration Fraction. This is the amount of plasma filtered from the glomerulus into the Nephron.

Eighty percent of the plasma volume remains in the blood stream. More than 19 % of the filtered blood gets reabsorbed back into the bloodstream. This leaves less than one percent of the filtered blood for excretion.

The afferent arterioles go into the renal corpuscle. It supplies each glomerulus with blood. Substances from the glomerulus then get filtered into the Nephron. This process is also known as Glomerular Filtration.

Mesangial cells occupy the space between the capillaries of the glomerulus. These cells have actin that causes them to contract and relax. Mesangial cells help to regulate blood flow.

Endothelial cells line the capillaries of the glomerulus. Endothelial cells contain many pores called fenestrae. These tiny pores allow solutes to move from the capillary into the nephron.

The glomerulus also has a basement membrane. There are spaces between the glomerular basement membrane called filtration slits. These filter water, glucose, and blood proteins.

Above the glomerular basement membrane are podocytes. These are feet like projections that have gaps which also allow for filtration. Endothelial cells, basement membrane and the podocyte make up the Filtration barrier.

Substances that are able to pass through the Filtration barrier become glomerular filtrate. The filtration barrier does not allow large molecules to pass through to the Nephron. It also doesn't allow negative charged substances.

The basement membrane and podocytes have a negative charge. This means that smaller and molecules with a positive charge can get absorbed. Proteins for example, have a negative charge and are too big to pass through the filtration slits.

But the basement membrane can get damaged by certain kidney diseases. This will allow protein to leak in to the filtration slits and into the urine.

Filtration pressure causes substances to move from the glomerulus into the nephron. Hydrostatic pressure (Pgh) is the pushing force that allows for glomerular filtration.

This pressure is about 60 millimeter of mercury (60 mmHg). There are also pressures that oppose filtration. These types of pressures are for facilitating substances going back into the glomerulus.

The first is the Bowman's capsule hydrostatic pressure (Pbc). The Pbc is about 16 mmHg. The other form of pressure that opposes filtration is Osmotic Glomerulus Pressure (Pgco). Pgco is about 34 mmHg. The pulling force made by proteins cause Osmotic pressure.

The Net Filtration pressure (NFP) is equal to outward pressure minus inward pressure: - NFP = Pgh − Pbc − Pgco. This results in a NFP of +10mmHg. This positive pressure direction favors filtration.

GFR refers to the Glomerular Filtration rate. GFR is the total volume of fluid that filters into the Bowman's capsule per minute. It is a good indicator of kidney function.

The average GFR for healthy adults is 125 ml/min. GFR = a Filtration constant − NFP. The filtration constant comes from the surface area of the glomerular capillary.

Altering the NFR will also alter GFR. An Increase or decrease of GFR can show problems with the kidney's function.

Obstructions on the afferent arteriole cause a decrease in Pgh. This also results in decreased GFR. When the afferent arteriole becomes dilated with blood, both Pgh and GFR increases. Resistance or vasoconstriction in the efferent arteriole increases both Pgh and GFR.

Bowman's capsule hydrostatic pressure (Pbc) exerts pressure of its own. This pushes back against the glomerulus. Increased Pbc leads to decreased GFR, while decreased Pbc increases GFR.

Kidney disease damages the basement membrane. This also results in decreased GFR.

The Reabsorption process

The renal tubule of the Nephron has several segments. These segments allow for the reabsorption of substances back in circulating blood.

Following the Bowman's capsule is the Proximal Convoluted Tubule (PCT). Sodium, glucose, amino acids, and potassium get reabsorbed in this segment. The PCT also reabsorbs urea, phosphate, and citrate.

After the PCT is the Loop of Henle (LOH). The LOH is a U shaped tube and consists of a descending and an ascending limb.

Water reabsorption takes places in the Descending limb of loop of Henle. The descending section is less permeable salt. But Sodium Chloride and potassium reabsorption occurs in the Ascending limb of LOH.

The thicker ascending limb of LOH connects back to the distal convoluted tubule (DCT). The DCT reabsorbs sodium, chloride, potassium, calcium, magnesium and bicarbonate. It also regulates pH.

The DCT then connects into the Collecting Duct. The Collecting Duct reabsorbs sodium, water, chloride, and urea. The renal tubules also have cuboidal epithelium cells within them. These cells help in the reabsorption of substances.

The Secretion process

Several segments of the nephron also allow for the secretion of other substances. The substances get secreted from the blood into the collecting duct.

The PCT can secrete organic acids such as creatinine and also many types of medications. The DCT allow for the secretion of Hydrogen ions and potassium.

The PCT secretes the hormone erythropoietin. This hormone stimulates red blood cell production. The kidney also secretes the enzyme/hormone Renin. This enzyme helps to regulate arterial blood pressure. A decrease in sodium and blood pressure triggers the secretion of renin.

There are two types of nephrons. The shorter cortical nephron is more prominent in the renal cortex (outer part of the kidney). But it does not go deep into the renal medulla (inner part of kidney). The longer Juxtamedullary nephrons penetrate deep into the medulla.

Vasa recta surround the long loop of Henle of this longer type of nephron. Vasa recta are straight capillaries in the medulla and lie parallel to the loop of Henle. Only juxtamedullary nephrons have the loop of Henle surrounded by vasa recta.

The Vasa recta form when Efferent arterioles leave the glomerulus of juxtamedullary nephrons. Efferent arterioles are outgoing blood vessels of the glomerulus. They carry filtered blood away from the glomerulus.

The vasa recta allow for the secretion and reabsorption of water along the long loop of Henle. The vasa recta also help maintain water balance and produce concentrated urine.

Cortical nephrons do not have vasa recta. But they also have capillaries that surround the renal tubules. Efferent arterioles also break up into capillaries. And this also occurs when Efferent arterioles leave the glomerulus of cortical nephrons.

The capillaries will then travel along cortical portions of renal tubules. They enable cortical nephrons to perform reabsorption and secretion of substances.

Chemical reactions also occur within tubule lumen. Lumen is the inside space of the renal tube. In the PCT, glucose and amino acids get reabsorbed back into circulation. Sodium gets reabsorbed in the cell and exchanged with potassium. PCT also regulates acid base balance.

The lumen tube of the PCT also secretes hydrogen. Hydrogen ions (H+) inside the lumen react with bicarbonate ions (CO_3) to form carbonic acid (H_2CO_3). The enzyme carbonic anhydrase converts carbonic acid to water and carbon dioxide.

Carbon dioxide can diffuse back to the cell and into circulation. The body becomes more acidic when carbon dioxide levels increase. Bicarbonate can get reabsorbed into circulation to increase the pH of blood.

Carbon dioxide can react with water in the cell. A reverse reaction then takes place. Carbon dioxide and water become carbonic acid. Carbonic acid converts back to bicarbonate and hydrogen ions.

Other electrolytes such as calcium ions can also get reabsorbed in the PCT. Electrolytes also get reabsorbed in the Ascending LOH. Sodium, chloride, and magnesium get absorbed in the DCT.

Channels in the DCT and Collecting duct exchange sodium and potassium. The hormone aldosterone stimulates sodium reabsorption.

An increase in circulating aldosterone causes more sodium reabsorption. This results in the reabsorption of water to increase blood pressure. More potassium also gets produced.

Aquaporins in the collecting duct allow for the reabsorption of water. When the body reabsorbs water, it will increase the Osmolality of urine. This means that solute concentration in urine also increases.

An antidiuretic hormone, vasopressin controls the number of Aquaporins in the collecting duct. More Aquaporins means more water in circulation.

Urine gets produced after reabsorption and secretion of substances in the renal tube. Urine consists of excess water and toxic nitrogenous waste. It may also contain metabolites and white blood cells.

Red blood cells in the urine may show damage to the glomerulus. Urea is the component of urine which provides a way for nitrogen removal from the body.

Urine processed by filtration, reabsorption, and secretion leaves the kidneys. It exits through the ureter and stored in the bladder. Micturition (urinating) removes urine from the urethra. The amount of urine represents only one percent of the total filtered plasma volume.

The kidney reabsorbs glucose but it is capable of producing it as well. This is also known as renal gluconeogenesis. In general, gluconeogenesis takes place in the liver.

Glucose production

Gluconeogenesis occurs during starvation, when people fast and engage in intense workouts. The liver and the kidney produce glucose in these cases to maintain blood glucose levels.

The renal cortex produces glucose from non-carbohydrate carbon substrates. These include glycerol and lactate.

Urine should not contain glucose because the kidneys are able to reabsorb it. People with diabetes have elevated blood glucose levels. The Proximal tubule (PCT) can only reabsorb a certain amount of glucose. Excess glucose gets deposited back into the urine.

Conclusion

Anatomy and Physiology is a basic foundation of learning for health professionals. Being knowledgeable in this field promotes the proper application and clinical decisions.

Aside from technical skills, one should have a piece of scientific knowledge. This will give you the confidence to give the patient the proper care they need.

This book was meant to help you get started. The next thing to do is to dive even deeper and find out more about the complexities of the human body.

Lightning Source UK Ltd.
Milton Keynes UK
UKHW010749060521
383241UK00003B/458